SPRINGER PROTOCOLS HANDBOOKS

For further volumes:
http://www.springer.com/series/8623

Springer Protocols Handbooks collects a diverse range of step-by-step laboratory methods and protocols from across the life and biomedical sciences. Each protocol is provided in the Springer Protocol format: readily-reproducible in a step-by-step fashion. Each protocol opens with an introductory overview, a list of the materials and reagents needed to complete the experiment, and is followed by a detailed procedure supported by a helpful notes section offering tips and tricks of the trade as well as troubleshooting advice. With a focus on large comprehensive protocol collections and an international authorship, Springer Protocols Handbooks are a valuable addition to the laboratory.

Advanced Analytical Techniques in Dairy Chemistry

Kamal Gandhi

Dairy Chemistry Division, National Dairy Research Institute, Karnal, India

Neelima Sharma

National Referral Center for Milk Quality and Safety, National Dairy Research Institute, Karnal, Haryana, India

Priyae Brath Gautam

Dairy Chemistry Division, National Dairy Research Institute, Karnal, Haryana, India

Rajan Sharma

Dairy Chemistry Division, National Dairy Research Institute, Karnal, Haryana, India

Bimlesh Mann

Dairy Chemistry Division, National Dairy Research Institute, Karnal, Haryana, India

Vanita Pandey

Quality and Basic Sciences, Indian Institute of Wheat and Barley Research, Karnal, Haryana, India

Springer

Kamal Gandhi
Dairy Chemistry Division
National Dairy Research Institute
Karnal, India

Priyae Brath Gautam
Dairy Chemistry Division
National Dairy Research Institute
Karnal, Haryana, India

Bimlesh Mann
Dairy Chemistry Division
National Dairy Research Institute
Karnal, Haryana, India

Neelima Sharma
National Referral Center for Milk Quality and Safety
National Dairy Research Institute
Karnal, Haryana, India

Rajan Sharma
Dairy Chemistry Division
National Dairy Research Institute
Karnal, Haryana, India

Vanita Pandey
Quality and Basic Sciences
Indian Institute of Wheat and Barley Research
Karnal, Haryana, India

ISSN 1949-2448 ISSN 1949-2456 (electronic)
Springer Protocols Handbooks
ISBN 978-1-0716-1939-1 ISBN 978-1-0716-1940-7 (eBook)
https://doi.org/10.1007/978-1-0716-1940-7

This Springer imprint is published by the registered company Springer Science+Business Media, LLC, part of Springer Nature.
The registered company address is: 1 New York Plaza, New York, NY 10004, U.S.A.

Contents

About the Authors ... *xiii*
Abbreviations ... *xv*

1 Basic Laboratory Skills ... 1
1 Laboratory Safety ... 1
2 Validation ... 2
3 Basic Tools and Operations ... 4
3.1 Electronic Weighing Balance ... 4
3.2 pH Strips and pH Meter ... 6
3.3 Volumetric Laboratory Equipment ... 7
3.4 Titration ... 8
4 Laboratory Waste Management ... 8
References ... 9

2 Chromatography ... 11
1 Chromatographic Parameters ... 11
2 Types of Chromatography ... 15
2.1 Paper Chromatography ... 16
2.2 Thin Layer Chromatography (TLC) ... 16
2.3 Column Chromatography ... 17
2.4 Adsorption Chromatography ... 18
2.5 Size Exclusion Chromatography (SEC) ... 18
2.6 Ion Exchange Chromatography (IEC) ... 21
2.7 Affinity Chromatography ... 22
2.8 Gas Chromatography ... 34
2.9 High-Performance Liquid Chromatography (HPLC) ... 57
References ... 83

3 Centrifugation ... 85
1 Principle ... 85
1.1 Buoyant Force ... 87
1.2 Frictional Force ... 87
1.3 Derivation of Stokes' Law ... 87
1.4 Sedimentation Coefficients ... 88
2 Classification of Centrifuges ... 88
2.1 Desktop Clinical Centrifuges ... 88
2.2 High-Speed Centrifuges ... 89
2.3 Microcentrifuge ... 90
2.4 Vacuum Centrifuge/Concentrators ... 90
2.5 Ultracentrifuge ... 90
2.6 Instrumentation of an ultracentrifuge: ... 92
3 Types of Rotors ... 93
3.1 Fixed Angle Rotors ... 93
3.2 Vertical Tube Rotors ... 93

3.3 Swinging-Bucket Rotors. 94
4 Preparative Centrifugation 94
4.1 Differential Gradient Centrifugation 95
4.2 Density Gradient Centrifugation 95
4.3 Analytical Centrifugation 98
5 Applications of Centrifuge. 100
6 Micellar Casein (MC) Preparation by Ultracentrifugation 101
References 102

4 Polyacrylamide Gel Electrophoresis 103
1 Introduction: Principle, Components, and Gel Media. 103
2 Sample Preparation and Buffer Systems. 105
3 Types of Gel Electrophoresis Commonly Used for Milk Protein Separation . . . 106
3.1 Native PAGE 106
3.2 Urea PAGE. 108
3.3 Sodium Dodecyl Sulfate–Polyacrylamide Gel Electrophoresis (SDS PAGE) 108
3.4 Tricine PAGE. 110
3.5 Isoelectric Focusing (IEF) 111
3.6 Two-Dimensional Gel Electrophoresis (2D-GE) 112
4 Visualization and Detectiondetection. 113
5 Chemistry of Milk Proteins Under Electrophoresis 114
6 Applications in Dairy Science 115
7 Tricine-SDS-PAGE Protocol 116
7.1 Reagents 116
7.2 Gel Preparation 117
7.3 Sample Preparation 117
7.4 Electrophoresis 117
7.5 Gel Staining 118
References 120

5 Western Blotting. 121
1 Introduction: Principle and Components 121
2 Components of Western Blotting 123
3 Visualization and Detection 125
3.1 Visualization of Protein Markers 125
3.2 Visualization of Proteins in the Samples 125
4 Applications in Dairy Science 129
References 129

6 Membrane Processes 131
1 Basic Principle. 131
2 Advantages of Membrane Separation Process. 133
3 Drawbacks of Membrane Technology 133
4 Principal Types of Membranes. 134
4.1 Isotropic Membrane 134
4.2 Anisotropic Membranes 135
4.3 Inorganic Membranes. 136

4.4 Gas Separation/Permeation 136
5 Membrane Processes.................... 136
5.1 Reverse Osmosis.................... 137
5.2 Nanofiltration 138
5.3 Ultrafiltration.................... 138
5.4 Microfiltration 138
5.5 Dialysis.................... 139
5.6 Gas Permeation.................... 139
5.7 Pervaporation.................... 140
5.8 Electrodialysis 141
5.9 Liquid Membranes.................... 141
6 Fractionation of Milk Proteins by Ultrafiltration.................... 143
6.1 Principle.................... 143
6.2 Materials 144
6.3 Procedure 144
References 145

7 Potentiometry 147
1 Principle.................... 148
2 Potentiometric Electrodes.................... 149
2.1 Metallic Electrodes.................... 149
2.2 Membrane Electrodes 149
3 pH Meter and Measurement of pH 154
4 Buffer Solutions.................... 157
4.1 pH of a Buffer 157
5 Measurement of pH of Milk and Whey Sample.................... 157
5.1 Principle.................... 157
5.2 Materials and Reagents 157
5.3 Guidelines to be Followed While Operating the pH/Ion Meter.................... 157
5.4 Standardization/Calibration of pH Meter 158
5.5 Measuring pH of the Sample 158
6 Applicability of ISE for Measuring the Ionic Calcium in Skim Milk.................... 158
6.1 Principle.................... 158
6.2 Preparation of Calibration Curve.................... 159
6.3 Materials and Reagent.................... 159
6.4 Procedure 159
References 160

8 Spectroscopy 161
1 Principle.................... 163
2 Beer's Law.................... 165
3 Limitations to Beer's Law 165
3.1 Fundamental Limitations.................... 166
3.2 Chemical Limitations 166
3.3 Instrumental Limitations.................... 166
4 Factors Influencing the Absorption Spectra of Chromophores 166
5 Energy Levels in Atoms and Molecules 167
6 Components of UV-Vis Spectrophotometer 168
6.1 Sources of Electromagnetic Radiation 168

6.2 Sample Holding Cell ... 170
6.3 Detectors ... 170
6.4 Signal Processors ... 171
7 Single-Beam vs Double-Beam Spectrophotometer ... 172
8 Fluorescence Spectroscopy ... 173
9 Determination of Absorption Spectrum of Bovine Serum Albumin (BSA) ... 174
9.1 Principle ... 174
9.2 Materials and Reagents ... 174
9.3 Procedure [5] ... 174
10 Verification of Beer's Law Using Bovine Serum Albumin (BSA) ... 174
10.1 Principle ... 174
10.2 Materials and Reagents ... 174
10.3 Procedure ... 175
11 Effect of pH on the λ_{max}, Absorbance (A) and Absorbtivity of p-Nitrophenol Solution [5] ... 175
11.1 Principle ... 175
11.2 Materials and Reagents ... 175
11.3 Procedure ... 175
11.4 Observations and Calculations ... 176
References ... 176

9 Infrared (IR) Spectroscopy ... 177
1 Infrared Region of the Electromagnetic Spectrum ... 177
2 Molecular Vibrations ... 179
2.1 Infrared Activity ... 179
3 Factors Affecting Absorption of Frequency ... 180
4 Regions of IR Spectra ... 181
4.1 The Fingerprint Region ... 181
4.2 Functional Group Region ... 181
5 Comparison of Spectra of Alkanes, Alkene, and Alkyne ... 182
6 Dispersive vs Fourier Transform Instruments ... 183
6.1 Dispersive Instruments ... 183
6.2 Fourier Transform Infrared Spectroscopy ... 184
6.3 Components of FTIR Instruments ... 185
6.4 Advantages of FTIR in Food Analysis ... 186
6.5 Advantages of FTIR over Dispersive IR ... 186
6.6 Sample Preparation and Handling Technique ... 186
6.7 Interpretation of the Spectra ... 188
7 Applications of Mid-IR Spectroscopy ... 189
7.1 Qualitative Applications ... 189
7.2 Quantitative Applications ... 189
7.3 Placement of IR Analyzer in a Dairy Industry ... 191
8 Attenuated Total Reflectance-Fourier Transform Infrared Spectroscopy (ATR-FTIR) ... 192
8.1 Basic Principle of ATR Spectroscopy ... 192
8.2 Designs of ATR ... 194
8.3 ATR Crystals Materials ... 194
8.4 Advantages of ATR IR Spectroscopy ... 195
8.5 Application of ATR IR Spectroscopy ... 196

9 Application of Chemometrics to Develop Spectroscopic Method 196
9.1 Selection of Variables 197
9.2 Calibration of Model 197
9.3 Review of Classification Results 197
9.4 Validation of the Model 197
9.5 Selection of Algorithm 197
9.6 Validation of Calibrated (Prediction) Model 197
References 198

10 Mass Spectroscopy 199
1 Principle of Mass Spectroscopy 200
2 Steps Involved in Mass Spectroscopy 200
3 Methods of Ionization 201
3.1 Matrix Assisted Laser Desorption Ionization (MALDI) 201
3.2 Electrospray Ionization 203
3.3 Atmospheric Pressure Chemical Ionization 204
3.4 Atmospheric Pressure Photoionization 205
4 Mass Analyzers 206
4.1 Types of Mass Analyzer 207
5 Advantages of MS 215
6 Acquisition Methods 215
References 217

11 Atomic Absorption Spectroscopy and Flame Photometry 219
1 Introduction 219
2 Basic Principles of AAS 220
2.1 Sample Atomization Steps 221
3 Processing of Samples 222
4 Flame Atomic Absorption Spectroscopy 224
4.1 Radiation Source 226
4.2 Atomizers 226
4.3 Burner 227
4.4 Monochromator 227
4.5 Detector 227
4.6 Readout Devices 227
4.7 Advantages of Flame AAS 227
4.8 Disadvantages of Flame AAS 228
5 Graphite Furnace (Electrothermal) Atomic Absorption Spectroscopy (GFAAS) 228
5.1 Advantages of Furnace AAS 229
5.2 Disadvantages of Furnace AAS 229
5.3 General Practical Considerations of AAS 229
5.4 Sample Preparation 229
5.5 Labwares 229
5.6 Calibration 230
5.7 Standard Addition Method 230
6 Interferences in Atomic Absorption Spectroscopy 230
6.1 Spectral Interference 230
6.2 Absorption of Source Radiation by Background 230
6.3 Nonspectral Interference 230

7 Hydride Generation AAS (HGAAS) 231
8 Mercury Cold Vapor Generation 232
9 Correction of Drift 232
9.1 Double-Beam Optics 232
9.2 Stockdale Optics 232
10 Correction of Background 233
10.1 Advantages 233
10.2 Disadvantages 233
10.3 Structured Background 233
10.4 Zeeman Effect 234
10.5 Disadvantages of Zeeman Correction 234
11 Inductively Coupled Plasma-Atomic Emission Spectroscopy (ICP-AES) 234
12 Inductively Coupled Plasma-Mass Spectrometry (ICP-MS) 235
12.1 Introduction of Sample 236
12.2 Working of RF Generator and ICP Torch 236
12.3 Reaction 236
12.4 Monochromator 236
12.5 Detector and Analysis Through Computer 236
13 Flame Photometry 236
13.1 Estimation 239
13.2 Application 239
13.3 Disadvantages 239
13.4 Preparation of Sample 240
13.5 Preparation of Standard Solution 240
13.6 Procedure for Preparation of Stock 240
13.7 Determination of Metals in the Unknown 240
13.8 Precautions during the Analysis 241
14 Flame Photometry vs AAS 241
15 Applications of These Techniques 242
15.1 Determination of Iron Content in Infant Milk Substitute: Atomic Absorption Spectrophotometric Method 242
15.2 Determination of Sodium and Potassium Content in Milk Samples by Flame Photometry 245
References 247

12 Lateral Flow Assay 249
1 Introduction 249
2 Components of Lateral Flow Assay 250
2.1 Sample Pad 250
2.2 Conjugate Pad 251
2.3 Membrane 251
2.4 Adsorbent Pad 251
2.5 Backing Card 252
2.6 Position of the Test Line 252
3 Principles or Formats of Lateral Flow Assay 252
3.1 Sandwich Format 252
3.2 Competitive Format 253

3.3 Multiplex Detection Format 254
3.4 Adsorption–Desorption Format 256
4 Biorecognition Molecules 256
4.1 Antibodies 256
4.2 Aptamers 257
4.3 Molecular Beacons 259
5 Labels for Detection 259
5.1 Gold Nanoparticles (AuNPs) 260
5.2 Enzymes 260
5.3 Colloidal Carbon 261
5.4 Colored Latex Beads 261
5.5 Magnetic Particles and Aggregates 261
5.6 Fluorescent and Luminescent Materials 261
6 Characterization of Nanoparticles 262
7 General Steps for Lateral Flow Assay 263
7.1 Antibody Development against Target Analyte 263
7.2 Preparation of AuNPs 264
7.3 Conjugation of AuNPs with Antibody 264
7.4 Construction of the LFA Strip and Analyte Detection 265
8 Application of Lateral Flow Assay in Dairy and Food 265
9 Advantages and Limitations of LFA 266
10 Conclusion 268
References 269

About the Authors

KAMAL GANDHI is a scientist in the Department of Dairy Chemistry, at the ICAR-National Dairy Research Institute, Karnal, India. He received his Ph.D. in Dairy Chemistry from National Dairy Research Institute University in 2014. He has work experience of one and half years in Gujarat Cooperative Milk Marketing Federation (GCMMF), Amul. His area of expertise includes milk and milk products adulteration detection, functional foods, and milk lipids. He has published over 30 research publications in national and international journals. He is a life member of the Indian Science Congress Association, Association of Food Scientists and Technologists, India (AFSTI), and the Indian Dairy Association (IDA). He is the recipient of Early Career Research Award (Project for three years) from Science and Engineering Research Board, Department of Science and Technology, Government of India.

NEELIMA SHARMA is a postdoctoral research scholar at National Referral Center for milk quality and safety-chemical section at the National Dairy Research Institute, Karnal, Haryana, India. She received her Ph.D. degree in Dairy Chemistry from National Dairy Research Institute University in 2013. Her specialization is in milk proteins and peptides.

PRIYAE BRATH GAUTAM is currently pursuing his PhD in Dairy Chemistry at the National Dairy Research Institute, Karnal. He was the Deputy Manager (Quality Assurance) of the Punjab State Co-operative Milk Producers' Federation Limited for 2 years.

RAJAN SHARMA is a Principal Scientist at the Department of Dairy Chemistry, ICAR-National Dairy Research Institute, Karnal, India. He has around 23 years of experience in the area of milk quality and analytical Dairy Chemistry. He has been associated with the Food Safety and Standards Authority of India (FSSAI) since 2009, as Member of Scientific Panel on Milk and Milk Products, as well as Methods of Sampling and Analysis. Presently, he is working as a Chairman of FSSAI Scientific Panel on Methods of Sampling and Analysis. He is also working with the National Accreditation Board for Testing and Calibration Laboratories (NABL) as empanelled assessor since 2003. Many of the rapid methods developed by his group for assessment of quality of milk have been commercialized to Dairy Industries. He is a recipient of the NRDC Meritorious Invention Award—2013 and has been conferred Fellowship of the National Academy of Agricultural Sciences (2018) and National Academy of Dairy Science (2014).

BIMLESH MANN is a Principal Scientist at the Department of Dairy Chemistry Division, ICAR-National Dairy Research Institute, Karnal, India. She served as Head, Department of Dairy Chemistry, ICAR-National Dairy Research Institute from 2014 to 2020. Her research over the last 30 years has focused on the chemistry of milk and milk products with an emphasis on bioactive milk proteins and peptides, functional dairy foods, and nano encapsulation of bioactive components for dairy foods. Apart from this, she is also involved in research related to quality assurance of dairy products. She is also associated with Food Safety and Standard Authority of India as member of Milk and Milk Product Panel since 2020. She is the recipient of Best Teacher Award from three different organizations: Indian Council of Agricultural Research (2014), ICAR-National Dairy Research Institute (2012), and Association of Food Scientists and Technologists (INDIA) (2013). She is the editor of *Indian Journal of Dairy Science* published by Indian Dairy Association.

VANITA PANDEY is a Scientist at the Indian Institute of Wheat and Barley Research, Karnal. She is a Gold Medalist for her PhD research work. Her area of expertise includes plant biochemistry, molecular biology, plant tissue culture, and enhancement of nutritional and processing quality of wheat.

Abbreviations

2DE	Two-dimensional electrophoresis
2D-GE	Two-dimensional gel electrophoresis
2ME	2-mercaptoethanol
AAS	Atomic absorption spectroscopy
AES	Atomic emission spectroscopy
API	Atmospheric pressure ionization
APPI	Atmospheric pressure chemical ionization
APPI	Atmospheric pressure photon ionization
APS	Ammonium persulfate
ATR-FTIR	Attenuated total reflectance-Fourier transform infrared spectrophotometer
AuNPs	Gold nanoparticles
BLM	Bulk liquid membrane
BSA	Bovine serum albumin
CAD	Collisionally activated dissociation
CAF	Chemically assisted fragmentation
CE	Collision energy
CID	Collision-induced dissociation
CM	Carboxymethyl
DDA	Data-dependent analysis
DEAE	Diethyl aminoethyl
DLS	Dynamic light scattering
DNA	Deoxyribonucleic acid
DP	Declustering potential
DT	Dwell time
DTT	Dithiothreitol
ECD	Electron capture dissociation
ECD	Electron capture detector
ELM	Emulsion liquid membrane
ESI	Electrospray ionization
ETD	Electron transfer dissociation
FAB	Fast atomic bombardment
FEP	Flame emission photometry
FID	Flame ionization detector
FT	Fourier transform
FTIR	Fourier transform infrared radiation
GC	Gas chromatography
GFAAS	Graphite furnace (electrothermal) atomic absorption spectroscopy
GLC	Gas–liquid chromatography
GMP	Glycomacropeptide
GPC	Gel permeation chromatography
HATR	Horizontal ATR
HCL	Hollow cathode lamp
HDPE	High-density polyethylene
HG	Hydride generation accessories
HGAAS	Hydride generation atomic absorption spectroscopy

HPLC	High pressure liquid chromatography
HRP	Horse radish peroxidase
ICP-AES	Inductively coupled plasma-atomic emission spectroscopy
ICP-MS	Inductively coupled plasma-mass spectrometry
ICP-OES	Inductively coupled plasma-optical emission spectroscopy
IEF	Isoelectric focusing
IR	Infrared
ISE	Ion selective electrode
KLH	Keyhole limpet hemocyanin
LC	Liquid chromatography
LFA	Lateral flow assay
LFIA	Lateral flow immunochromatographic assay
LIT	Linear ion trap
MALDI	Matrix-assisted laser desorption ionization
MB-ATR	Multiple bounce ATR
MC	Micellar casein
MP-AES	Microwave plasma-atomic emission spectroscopy
MS	Mass spectrometry
MSA	Magnetic sector analyzer
NHS	N-hydroxysuccinimide
NPD	Nitrogen phosphorus detector
NTA	Nanoparticles tracking analysis
OD	Optical density
OPD	Optical path difference
PA	Polyacrylamide
PAGE	Polyacrylamide gel electrophoresis
PAGs	Polyacrylamide gels
PAS	Periodic acid Schiff
PBS	Phosphate buffered saline
pI	Isoelectric point
PMT	Photomultiplier tube
POC	Point of care
PVDF	Polyvinylidene fluoride
QDs	Quantum dots
RCF	Relative centrifugal force
RI	Refractive index
RMRD	Raw Milk Reception Dock
RNA	Ribonucleic acid
RO	Reverse osmosis
RP-HPLC	Reverse phase high performance liquid chromatography
RPM	Revolutions per minute
SB-ATR	Single bounce ATR
SCOT	Support coated tubular columns
SDS	Sodium dodecyl sulfate
SDS-PAGE	Sodium dodecyl sulfate-polyacrylamide gel electrophoresis
SEC	Size exclusion chromatography
SELEX	Systematic evolution of ligands by exponential enrichment
SLM	Supported liquid membrane

TBS	Tris-HCl buffer saline
TCD	Thermal conductivity detector
TEA	Triethylamine
TEMED	Tetramethylenediamine
TFA	Trifluoroacetic acid
TLC	Thin layer chromatography
TOF	Time of flight
UCNPs	Upconverting phosphorous
UV	Ultraviolet
VGA	Vapor generation accessories
WCOT	Wall-coated open tubular

Check for updates

Chapter 1

Basic Laboratory Skills

Abstract

Knowledge of basic laboratory skills can pay huge dividend in terms of overall health of the pupil as well as accuracy of results by decreasing chances of errors and accidents while working in the laboratory. This chapter outlines general safety measures to be followed by analysts and researchers while performing any experiment. Important signs/symbols used in most of the laboratories are also given in pictorial form. Further, commonly used tools and operations, namely, electronic weighing balance, pH strips, pH meter, volumetric equipment, and titration, are also explained in detail. Lastly, management of laboratory waste (especially chemical section) required for the overall safety of the lab is also briefly explained in this section.

Keywords Validation, Aqua regia, Weighing balance, Titration, pH meter

1 Laboratory Safety

Safety while working in a laboratory is crucial. For this reason, newcomers in the lab are often given formal lab orientation and safety training to make them aware about the areas where particular safety measures need to be taken. Similarly, standard operating procedures (SOPs) for equipment and other analysis routinely practiced safely in the lab are either attached to the equipment itself or kept near the area where the analysis is usually done. All these things are done to make sure that chances of error and accidents are reduced to the least possible level. Though detailed understanding regarding the unique safety measures being undertaken in a particular lab can be acquired only after demonstration by a trained instructor working in that lab, but in general, the safety measures which are to be followed while working in the lab are discussed below:

- While working in the lab one should always wear a lab coat, safety eyeglasses, close-toe shoes, minimum jewelry and tie back long hair. Appropriate gloves (depending on the type of analysis) should be worn and changed regularly. Touching of face, eyes, nose, ears, etc. in the laboratory should be avoided.

Kamal Gandhi et al., *Advanced Analytical Techniques in Dairy Chemistry*, Springer Protocols Handbooks,
https://doi.org/10.1007/978-1-0716-1940-7_1,

- Make oneself aware of the safety exit points, safety showers, firefighting equipment, fire alarms, and first aid locations.
- Eating, drinking, and smoking in the lab should be avoided.
- Before leaving lab, one should ensure that all the equipment, lights, air conditioner, etc. are turned off or appropriately kept as per SOPs if running overnight.
- One should be familiar with the signs commonly found in most labs (Fig. 1).
- Acids, especially concentrated, can cause body tissue damage and pain. While skin contact with bases often goes unnoticed because they do not cause immediate pain; however, they are corrosive and can damage body tissue. Eyes are particularly sensitive to both acids and bases, in case of any eye irritation, they should be thoroughly washed with water.
- Appropriate cleaning up of spill as soon as possible is important. Never clean up or neutralize acid spills with bases and vice versa. A potential exothermic reaction may occur which can be aggressive. Instead, commercially available spill-up cleaning kits for acids and bases can be used. These neutralize the spilled acid or base in a controlled manner without much evolution of heat.
- Always keep reactive and highly corrosive reagents like aqua-regia in fume hood.
- Dilution of acids/bases is exothermic. Therefore, always add them slowly in water. Never add water to acid/base. Further, acid drops can eat away clothes. This might not occur instantly rather small holes appear after washing.

2 Validation

Validation is a process of proving that the method used for analysis gives accurate results/data for the intended application. Broadly, it consists of two steps: (a) **Knowing what the problem is**—when the analyst poses the correct question and knows the data requirements then the overall process benefits and (b) **Knowing how it can be resolved**—when the analyst has adequate knowledge of method of analysis for the intended purpose.

In order to validate a method one needs to include studies on calculating the following:

- Selectivity: It is the index to measure that the method can detect the analyte of interest without interference from other substances.
- Linearity: It is the measure of increase in response in proportion to that of the addition of analyte. Generally, five to seven concentrations are studied to form the standard curve and then concentration of the unknown solution is calculated using

Fig. 1 Important signs normally seen in a laboratory

regression equation. Linearity is judged from the R^2 (coefficient of determination) value. Correlation coefficient, y-intercept, slope of regression line, and residual sum of squares should be presented together with the slope of data.

- Accuracy: It can be determined by recovery method, comparison with a standard method or analyzing reference material. It is the measure of closeness of the estimated value with the known value. When applied to a set of results, it involves a combination of random error and common systematic error.
- Precision: The closeness of agreement between independent analytical results obtained by applying the experimental procedure under the stipulated conditions. The smaller the random part of the experimental errors which affect the results. It is expressed as %RSD for a statistically number of samples. Repeatability expresses the precision (spread of the data, variability) under the same operating conditions over a short interval of time. It is termed as intra-assay precision. Intermediate precision expresses within-laboratories variations (different days, different analysts, different equipment). Reproducibility expresses the precision between the laboratories.
- Range: For a particular method, the working range is the one which gives accurate and precise results.
- Limit of detection: It is the lowest concentration of an analyte in a sample that can be detected, not quantified. In general, ten times the mean or three times the standard deviation of the sample blank is taken as limit of detection.
- Limit of quantification: It is the lowest value that can be measured in the sample with reasonable degree of accuracy and precision under stated operational conditions. It is generally taken as the value which is ten times the standard deviation above the sample blank.
- Robustness: When the estimated values are reliable, irrespective of different analyses, source of reagents, laboratories, instruments, etc.

3 Basic Tools and Operations

Though a detailed understanding of laboratory tools and operations can only be acquired when a trained instructor demonstrates them, a brief discussion about their use is discussed in the following sections.

3.1 Electronic Weighing Balance

These have replaced the previously used mechanical weighing balances because of their convenience and relatively less chances of errors. In general, these are available in different ranges and can perform weighing up to four to six decimal places. Modern electronic balances work on the principle of electromagnetic force restoration. Here the weighing pan is connected to an electromagnetic coil (through which current flows) around which magnetic field is created by an amplifier. The amplifier maintains the correct current to keep the lever in position. As more weight is placed over the pan, the amplifier creates more current to counter it. The counteracting force is electronically translated into digital signals, and numbers appear on the display. Some of the important points to be taken care of while using the balance are: (a) Location of the balance should be such that there are no vibrations from other equipment. (b) The "surface level bubble" tool should be used to bring the four legs of the balance at level. One should check whether the bubble is at the center or not every time prior to weighing. (c) Another important aspect is cleanliness; any residual material should be removed before use and spills (in and around the balance) should be removed immediately to avoid any damage (like corrosion) to the equipment. To remove the dust/powder from weighing balance use tissue or appropriate size brushes. Never blow the spill as it may go inside the balance. Use 70% isopropyl alcohol or ethanol sprayed on tissue to wipe away sticky substances (never pour them directly on the balance). Further balances should never be tilted or dislocated while cleaning. (d) Balances should be kept in draught-free locations, that is, away from windows, doors, fans, air conditioners, etc. to avoid fluctuations in the readings. In order to decrease the effect of draughts, many digital balances these days are equipped with doors to shield the weighing plates. (e) Balance should be kept away from those equipment which generate strong magnetic field. It can lead to permanent damages in the balance by affecting the response. (f) Regular calibration (generally annually) by trained personnel should be done and sticker of calibration date should be attached to the balance itself. Apart from external calibration, intermediate check should be done in-house at predetermined intervals by trained personnel using calibrated weight box. (g) Frequent accuracy check (daily, weekly, or monthly) by placing known weights (using forceps) to check their weight on display is recommended to make sure proper working of the equipment. (h) Leaving balance in stand-by mode is recommended instead of switching on and off frequently [1, 2].

There are two methods of weighing, that is, (a) by weighing after taring the weight of weighing vessel and (b) weighing by difference in which first weight of sample plus weighing vessel (W_1) is taken and then weight of empty vessel (W_2) is subtracted to get the weight of sample ($W_1 - W_2$). General protocol for use of electronic weighing balance is shown in Fig. 2. Further, one of the common problems that come across while using this instrument is "unstable reading." This is mainly due to lack of initial warming up leading to a thermal gradient between the sample vessel and

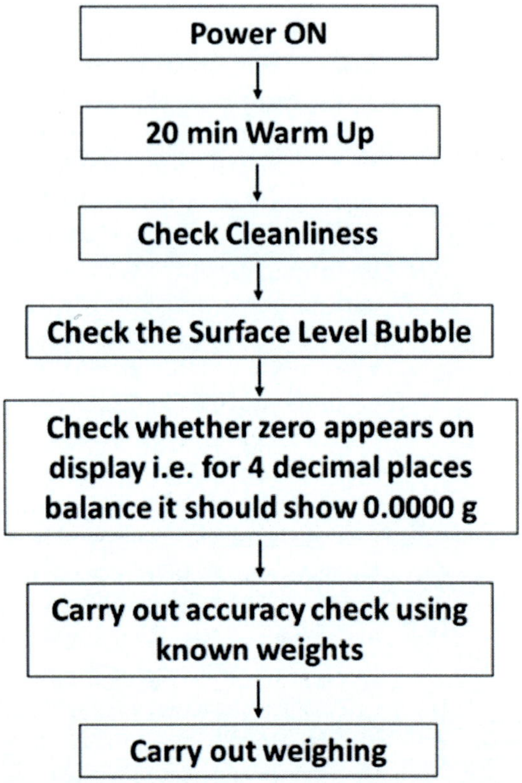

Fig. 2 Schematic diagram representing the use of electronic weighing balance

environment. Another reason can be sample temperature (cold objects appear heavier than warm one), volatility and hygroscopicity leading to variable readings until equilibrium is reached.

3.2 pH Strips and pH Meter

These are used to determine pH. The pH is a measurement unit that indicates acidic or alkaline nature of a solution. It is measured in the range of 0–14. Zero being very acidic, 7 means neutral, and 14 means very alkaline. When the hydrogen ion concentration $[H^+] >$ hydroxyl ion concentration $[OH^-]$ the solution is acidic, when $[H^+] = [OH^-]$ it is neutral, and when $[H^+] < [OH^-]$ the solution is said to be alkaline. By definition, pH is the negative log of hydrogen ion concentration and the change in one unit of pH corresponds to tenfold change in $[H^+]$. The paper strips of different ranges are frequently used for quick measurement of pH (Fig. 3a). These are very handy and convenient to use. The pH strips have chemical compounds mounted over it which after dipping can undergo color change depending on the pH of the sample. The color can be matched with the reference color chart (generally while the strip is still wet) provided by the manufacturer with the product. While the results can be obtained instantly, the estimated pH is not accurate because of the subjectivity involved and therefore error of 0.3–1.0 pH unit can occur.

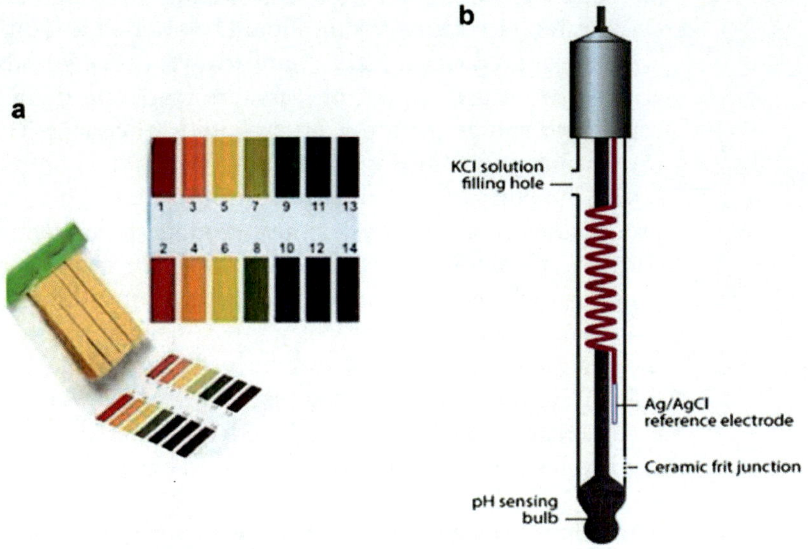

Fig. 3 (**a**) Paper based pH strips and (**b**) a combination electrode of pH meter

Conversely, pH meters can be used to accurately measure the pH of the solution. These are quite sensitive and capable of measurements between 0.01 and 0.1 pH units. The modern pH meters have two components: a sensing combination electrode (reference and measuring electrodes are mounted into same device) and high impedance pH meter. Inside the combination electrode, a solution having fixed pH is present which surrounds the reference electrode. When the combination electrode (Fig. 3b) is dipped in the solution (whose pH is to be measured), a potential is developed due to the difference in the concentration of $[H^+]$ in the sample and the solution inside the electrode. The pH meter reads this minute difference in voltage and electronically converts the signal to pH reading which finally appears on the display. While estimating the pH, one thing should be kept in mind that the temperature variation can bring a huge difference in pH values. Therefore, sample temperature should be brought to equilibrium before taking readings. Further, advanced electrodes known as automated temperature compensation electrode (ATCE) are available which can sense the temperature and give temperature corrected pH values.

3.3 Volumetric Laboratory Equipment

Accuracy of an analytical procedure highly depends on the accurate preparation of solutions/reagents to be used especially when quantitative determination has to be done. Certified glassware should be used for procedures where high accuracy is needed. Generally, glassware of two grades are available: Class A and Class B, the former having higher degree of accuracy with less measurement error. Further, proper care should be taken for cleaning of glassware. For washing, glassware should first be rinsed with the solvent/diluent previously used and afterwards with appropriate laboratory glassware detergent followed by thorough rinsing with clean water. Hot water rinsing should be avoided as it can affect the accuracy of graduation marks. Similarly, very strong acids or alkalis are likely to cause etching of glassware rendering them more susceptible to contamination. Certain liquids when mixed can lead to exothermic effect (mixing of methanol and water) or endothermic effect (mixing of methanol and acetonitrile). This can lead to production or absorption of heat leading to either increase or decrease of volume. Therefore, care should be taken that wherever possible such liquids should be separately measured and mixed afterwards. In general, for every 10 °C change in temperature, a measurement error of 1% can be expected. While reading the volume on the volumetric flask, burette, etc. care should be taken that meniscus should be read at eye level; otherwise, parallax error can happen.

3.4 Titration

Titration is determination of unknown concentration of an analyte by addition of a known concentration of a reagent. The accuracy of the titration depends on four factors, namely completeness of reaction, unambiguity of reaction, fast reaction, and ease in observation of end point. Completion of the reaction can be observed either via color change, pH meter, or electrochemical sensor. Although there are many types of titrations, one of the most common is acid–base titration. This is a quick and cost-effective titration in which an indicator (generally a dye) is added prior to titration and the end point is color change which can be observed either visually or using pH meter. The most common error is the tendency to get deeper color change to get permanent color past end point. Other types of titrations are: redox titration (indicator not used), complexometric (using EDTA), precipitation (using silver chloride), and zeta potential (for colloids) titration. Titration during the determination of the acidity in milk should be done comparatively fast as slow titration can result in conversion of ionic calcium to colloidal calcium phosphate and also result in fading of the phenolphthalein end point.

4 Laboratory Waste Management

The common wastes generated in a laboratory (especially in chemical section of analytical laboratory) are represented in Fig. 4. Many hazardous chemicals are frequently used in the laboratory. Proper care should be taken to dispose of each kind of waste as any carelessness can be hazardous. If not disposed off properly, these chemicals can gain entry into the environment and may also affect the people working nearby. For the better disposal of chemicals, it is recommended that each category should be collected separately, that is, organic solvents, acids, alkalies, explosive chemicals,

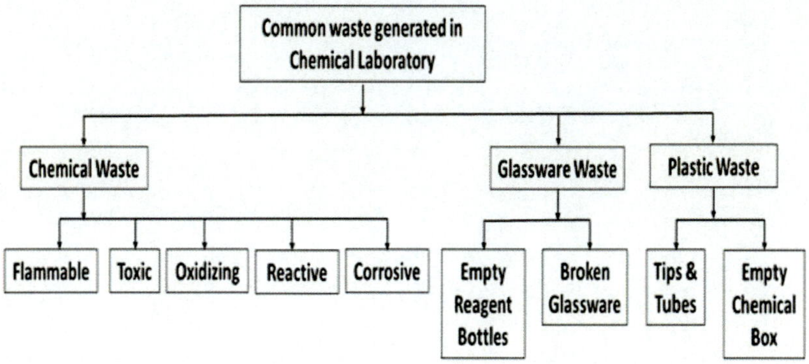

Fig. 4 Common wastes generated in a chemical laboratory

peroxide forming chemicals, toxic and carcinogenic chemicals, etc. in specific containers or biohazard bags. One should never mix the chemicals, for example, flammable chemicals should never be kept near oxidizers. Proper labeling of the container should be done so that the person responsible for garbage disposal can identify and follow proper disposal protocols. Likewise, there should be a separate dustbin for disposing of glassware waste. Broken glasses when disposed with other laboratory waste can be hazardous to the cleaning staff. A large portion of laboratory waste nowadays accounts to the plastic waste owing to the fact that plastic is replacing glass in almost every field. As plastic is a recyclable item, it should also be discarded separately. Flammable chemicals can be solids, liquids, or gaseous which get ignited when come in contact with flame, heat, or spark. There is a difference between flammable and combustible substances. While the former can readily burn at room temperature, the latter can burn after exposure to heat. One of the most common errors which a novice laboratory personnel does is to heat flammable chemicals (especially organic solvents during distillation process) using Bunsen burner instead of a water bath. This more than often becomes the reason for fire in the lab. Distillation experiments should be done in fume hood using water bath. Oxidizers can support ignition by supplying elements like oxygen and chlorine, thereby increasing the chances and intensity of fire. Oxidizers can also cause irritation of eyes, skin, and breathing passage. Similarly, toxic materials can lead to acute and chronic health effects [3].

References

1. https://www.chromacademy.com/
2. Christian GD (2007) Analytical chemistry. Wiley, Hoboken, NJ, USA
3. Chemical safety manual, Indian Institute of Technology, Bombay. Retrieved June 4, 2021, from. https://docplayer.net/35136643-Chemical-safety-manual.html

Chapter 2

Chromatography

Abstract

This chapter covers the principle, working, and operations of different types of chromatographic techniques like paper, thin layer, column chromatography, adsorption chromatography, size exclusion chromatography (SEC), ion exchange chromatography (IEC), affinity chromatography, high-performance liquid chromatography (HPLC), and Gas chromatography. Components of HPLC and GLC have been discussed in detail. The working and principle of detectors used in these techniques have been elaborated. Finally the applications of these techniques in the dairy chemistry have been discussed. Simple and validated protocols for the isolation and detection of amino acids by paper chromatography and thin layer chromatography, separation of milk fat components by thin layer chromatography, recovery of proteins from cheese whey using gel filtration, concentration of the dilute protein solution using gel filtration technique, separation of bovine serum albumin (BSA) and blue dextran using gel filtration and determination of K_{av} of bovine serum albumin, fractionation of casein by anion exchange chromatography, lactoferrin isolation from colostral whey using affinity chromatography with immobilized metal chelates, GC-FID analysis of the fatty acid content of milk fat, and determination of fatty acids of ghee using GC-MS/MS have also been covered.

Keywords Fatty acid, Paper chromatography, Thin layer chromatography, Column chromatography, Adsorption chromatography, Size exclusion chromatography (SEC), Ion exchange chromatography (IEC), Affinity chromatography, High-performance liquid chromatography (HPLC), Gas chromatography

Chromatography is a technique which facilitates the separation, purification, pre-concentration, and analysis of multiple analytes on an industrial and preparatory scale. Chromatographic methods are based on sample separation or distribution (of solute) between the stationary and moving phase. The relative migration of the solute between the two phases is the proportion of the concentration of the solute in stationary phase to that in the mobile phase described by partition or distribution coefficient (K_d).

1 Chromatographic Parameters

The ultimate aim in any chromatographic technique is to achieve the separation of all the components present in the sample mixture in minimum time. The three fundamental parameters, that is,

Kamal Gandhi et al., *Advanced Analytical Techniques in Dairy Chemistry*, Springer Protocols Handbooks,
https://doi.org/10.1007/978-1-0716-1940-7_2,

retention, selectivity, and efficiency must be taken into consideration so as to attain proper separation. These terms are related by the following equation:

$$R_s = 1/4\sqrt{N}\left(\frac{a-1}{a}\right)\left(\frac{k'}{k'+1}\right) \quad (1)$$

Among the three factors, selectivity has the greatest impact on improving the resolution. To ensure that the analyte(s) are baseline separated, the resolution value between the two peaks should be a minimum of 1.5. Resolution is given by the formula:

$$R_s = 2\Delta t/(w_2 + w_1) \quad (2)$$

where R_s = resolution.

Δt = difference between retention time of peak 1 and peak 2.

w_2 and w_1 = width of peak 2 and peak 1 at baseline.

Capacity factor or retention factor (*k*) gives a measure of the retention of the analyte. It is defined as the ratio of retention time of the analyte to the retention time of an unretained compound (Fig. 1). In other words, it can be defined as the velocity of the analyte relative to the velocity of the mobile phase. "*k*" is independent of flow rate and the column dimensions, while a high value of *k* indicates that the compound is strongly retained. So, the farther peaks in a chromatograph have higher retention factor, while the initial peaks will have a lower retention factor. Ideally retention factor ranges between values of 2 and 10. The analyte which does not get retained has no affinity for the stationary phase, so it elutes with the solvent at a time denoted by t_0 (dead time or hold up time). t_0 can be determined by several ways like including the time at the baseline disturbance observed due to the difference in the absorbance or the refractive index as the solvent passes through the injector. The most convenient approach is measurement of the peak width at half the height of the base peak which ensures that any problem associated with the tailing or non-baseline resolved peaks can be resolved. If the k value is less than 1, it means that the analyte is not going to be well retained in the column and it is going to elute very quickly from column which can lead to less stable separations. There are larger chances for the chromatographic interferences at the beginning of the chromatogram. At this point, even minute changes in the composition of the mobile phase can lead to changes in the retention. While in case of UHPLC (Ultra High Pressure Liquid Chromatography), retention factor values of less than one are usually obtained and they are less likely to be affected by any of these interferences occurring at the start of the chromatogram or by altering the mobile phase. This is due to the inherent efficiency of the technique. We sometimes require *k* value of greater that 10 when we have got a large number of components to be

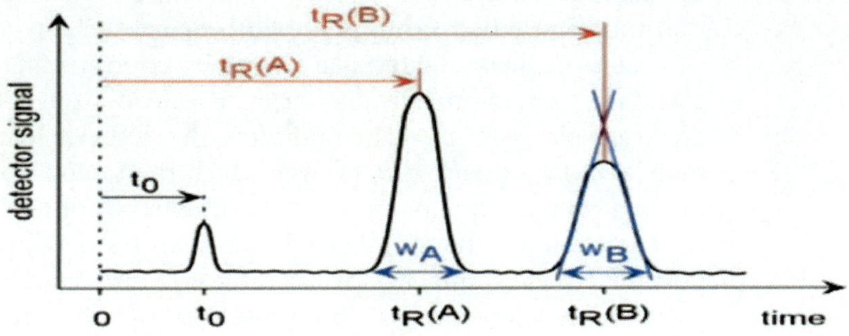

Fig. 1 Definition of retention time, t_R, and peak width, *w* [6, 7]

separated and resolve. But for higher *k* value, analyte retention in the column will be longer. Hence, broad peaks appear and baseline resolution decreases. Although, the resolution from half-height of the peak can be calculated, but if the broadening is incredibly severe, the peak might be lost altogether. For this reason, the *k* value is kept between 2 and 10 for the analyte in the sample. By altering the strength of the mobile phase, the value of *k* can be altered, and the largest gain in the separation is achieved with the *k* value ranging from 2 to 5.

The term **selectivity (*α*)** is the measure or ability of the chromatographic system to chemically distinguish between sample components and is calculated as the ratio of the retention factors (Fig. 1) as follows:

$$\alpha = \frac{k'_B}{k'_A} = \frac{t_R(B) - t_0}{t_R(A) - t_0} \tag{3}$$

the value α should be greater than one, a value equal to one indicates the co-eluting peaks. Higher the values of α indicate better separation efficiency power with better separation between the peak apices. We can alter the selectivity and separation by changing the type of the organic solvent. So, changing from acetonitrile to methanol or by adjusting the pH will change the ionization of our analyte and how it interacts with the mobile or the stationary phase. We can also change the solvent strength as well as the stationary phase chemistry. Temperature can also influence the selectivity. It has been already mentioned that the selectivity has the largest impact on the resolution. So, adjusting this parameter can have maximum gain in the resolution.

Efficiency (*N*) is defined as the extent to which the analyte gets dispersed when it travels through the column. This indicates the extent of band broadening as the analyte moves through the column and ideally the chromatographic peaks will be pencil thin lines, while the dispersion effects give rise to Gaussian shape peaks. Efficiency or *N* is generally referred as the plate number and a higher value of *N* is seen for the subsequent peaks in the chromatogram. Efficiency can also be measured by using the retention time and either using peak width or half-height peak width. N is derived from Martyn and Synge's comparison of column efficiency to fractional distillation, which divides the column into fictitious plates. Each plate indicates the distance between the mobile and stationary phases over which the sample components accomplish one equilibration. As a consequence, the effectiveness of separation and the number of equilibrations obtained are proportional to the number of plates in the column. Band broadening for column length L may be written as the theoretical plate's equivalent height, H, or as plate numbers, N (Eqs. 4 and 5). The chromatographic efficiency varies directly with the number of plates N and inversely to the H.

Plate number

$$N = 16\left(\frac{t_R}{w}\right)^2 \tag{4}$$

Plate height

$$H = \frac{1}{N} \tag{5}$$

There are several factors that negatively affect the efficiency: column, particle size of the packing column dimensions, injection volume, dead volume within the system, and flow rates. Typical plate number for a 5 μ column having dimensions of 4.6×100 mm is 5000–8000, higher is the number of plates in the column, lesser will be dispersion of the chromatographic bands. So, the less broad the peaks, the more efficient they will be, the narrower the peaks, greater is the impact on the resolution because we will have nice narrow peaks and these can base line separated easily.

Finally, there is one more parameter that is not linked through the resolution which is the **peak symmetry**. As mentioned, peak will be pencil thin lines, but they take Gaussian shapes due to dispersion, so we want our peaks to symmetrical. There should not be any fronting or trialing which makes the peak of strange shapes and might make them harder to resolve for the neighboring peaks. Some peaks will exhibit tailing and which can be caused by the dead volume in the system and column packing material (Fig. 2). Analytes can undergo secondary interaction with the silanol in the column which causes them to tail and to stick the column in a different way. To calculate the peak symmetry, peaks are split into two parts say A and B. Then we measure the distance at 10% of the peak height and we take the ratio of B over A. Ideally, we want values between 1 and 1.5, so that the symmetrical Gaussian peaks can be well baseline resolved. Fronting peaks (Fig. 2) can be caused if the concentration of the sample is too high or when the column is damaged or contains channels. Asymmetrical peaks often pose issues with chromatogram resolution and quantification of

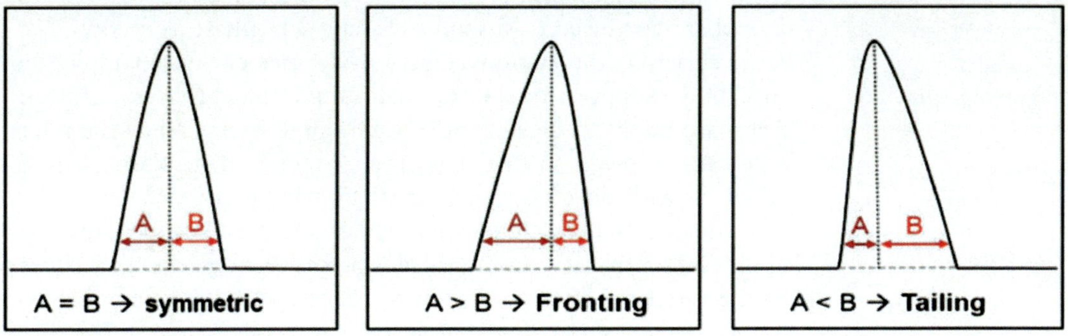

Fig. 2 Types of peaks. [12]

peaks. They are more difficult to overcome, and the integration of the peak to have a quantitative value is therefore much less reproducible.

2 Types of Chromatography

Chromatographic methods can be classified according to the nature of the mobile phases involved as in liquid chromatography (LC) and gas chromatography (GC). Only heat-stable gases or volatile liquids can be added to gas chromatography. A neutral carrier gas like oxygen, hydrogen, and helium is pumped into a heated column. Analysis of biomolecules like peptides and proteins cannot be analyzed by GC because of their higher molecular weight, they might get thermally decomposed before evaporation; however, smaller compounds like amino acids, carbohydrates, fatty acids can be analyzed using GC on their modification which enhances their thermal stability. GC can also easily analyze volatile metabolites like aldehydes, alcohols, or cell culture ketones. The sample is prepared and injected into a column filled with stationary phase. Because LC is not confined to volatile or heat-resistant chemicals, it is more flexible than GC [7–9]. The primary criterion for LC is that the analyte is soluble in the mobile phase. The refractive index, mass spectrometry, fluorescence, ultraviolet spectroscopy, and conductivity are all common techniques of detection. Normal-phase chromatography and reverse-phase chromatography are the two operational modes [14]. In normal-phase chromatography, the stationary phase is formed of polar or hydrophilic materials such as silica, while the mobile phase is formed of nonpolar or hydrophobic materials such as hexane. On the contrary, in reverse-phase chromatography, the stationary phase is more nonpolar than the mobile phase. The reversed phase chromatography, therefore, would be the ideal technique for the separating organic compounds like carbohydrates, nucleic acids, proteins, peptides, and amino acids. Chromatographic techniques can be classified broadly by the type of support used to sustain the stationary phase as follows.

2.1 Paper Chromatography

Paper serves as the stationary phase in paper chromatography by providing a support for the stationary fluid phase, that is, partition chromatography. A small spot of the sample is applied to the filter paper and allowed to dry. The dried paper is then placed in the sealed jar with developing solvent. The wick is prepared from the same material of the filter paper and is attached to the filter paper via a hole so that the developing solvent passes through the filter paper via the capillary action. The solvent then separates the sample into various components. Once the solvent passes across the entire paper, it is removed from the closed chamber and the component identification or sample identification is done by an appropriate method.

For the identification or visualization of isolated substances, physical, chemical, or biological methods may be applied. Physical procedures can involve electromagnetic radiation adsorption or emissions directly measured by the detectors. The chemical reaction requires pre- or post-derivatization of the separated compound that can be used before or after chromatography. Enzyme tests can be considered as an example of biological detection.

Water is used as the stationary phase in paper partition chromatography. On the contrary, the support may be impregnated with a nonpolar organic solvent and then developed in a polar solvent or water (reverse-phase paper chromatography). For complex sample mixtures, the sample is spotted in one corner of a square sheet of paper and developed in a single direction with a single solvent. After development of the chromatogram, it is rotated 90° and reconstructed using a second polarity liquid. The partition coefficient indicates the degree to which a material migrates. The distance traveled from the zone's center (d) to the developer's fronts is expressed in terms of the R_f value (D):

$$R_f = \frac{d}{D} \tag{6}$$

For a given solute/solvent/paper system, R_f values are not necessarily constant and these are not uniform parameters. The geometry of the solvent reservoir, direction of development, and the temperature also influence R_f values.

2.2 Thin Layer Chromatography (TLC)

This technique was described in 1938 and substituted paper chromatography because of the following reasons:

1. Rapid, more reproducible, and flexible.
2. Separation in paper chromatography is solely based on partitioning, but in TLC, it is reliant on the kind of chromatographic media utilized.

3. TLC experimental parameters can be conveniently modified to isolate and can be expanded for use in column chromatography.
4. Identification of a specific compound cannot be done using corrosive agents such as H_2SO_4 or elevated temperatures in paper chromatography.

This technique uses a small (250 μm thick) film of support material on a glass plate or on a commercially prepared surface. The chromatographic media can contain a binding agent such as $CaSO_4$ or gypsum to promote strong adhesion to the surface. After addition of the samples as a spot on the TLC plate, it is placed in a closed chamber such that the applied sample spot on the plate is nearer to the solvent. Through capillary operation, the liquid migrates upwards and the sample components get isolated. These isolated spots are visible by correct method after withdrawing the TLC plate from the forming chamber and evaporating the solvent. High-performance TLC is a modern comparatively high-performance technique (analogous to HPLC) where TLC plates are filled with tiny, more uniformly regulated porous particles. This enables greater separation in a shorter time. Both the normal and reverse phase chromatography have been used to detect adulterants (soyabeen oil and buffalo body fat) in ghee [3, 4].

2.3 Column Chromatography

In this technique, the stationary phases are water insoluble, porous, rigid particles. The size and shape of the stationary phase influence the flow rate and resolution characteristics. Big and coarse particles give poor resolution despite a faster flow rate while their counterparts exhibit better resolution efficiency in spite of the presence of smaller particles. The sample is placed on top of the column and eluted with a sufficient buffer for the fractionation and isolation of components. The eluate from the column is collected either by the automatic fraction collector or manually as fractions of the fixed volume or fractions eluted at a fixed retention time in separate tubes.

Analysis of the fractions is then carried out for the presence of the derived substance(s). The detection of the compound depends on its inherent properties like chemical, physical, or biological. Colored compounds can be viewed directly but for colorless compounds, other strategies are followed like their ability to give colored reactions with some other chemicals or on the basis of its physical properties like UV absorption, fluorescence, RI, or biological activity like enzymatic activity. This technique is classified on the basis of the type of interaction which occurs between the stationary phase and the sample or solute. Different types of the column chromatographic techniques have been discussed in the coming sections.

2.4 Adsorption Chromatography

Adsorption is a process in which molecules adhere to the surface of a strong adsorbent, forming distinct adsorption sites as a result of weak nonionic interactions such as Vander wall forces and hydrogen bonding. The strength with which the compound binds to the adsorbent varies and it can be desorbed selectively. So, selection of the right mobile phase and the adsorbent is an important factor to achieve good resolution. Alumina, charcoal, hydroxyapatite, silica, etc. are one of the most commonly used adsorbents. Polarity of the mobile phase is inversely related to the adsorption and its polarity also affects the adsorption process. Polar solvents are used when the sample or the solute has hydrophilic or polar groups; however, nonpolar or organic solvents are used when nonpolar or hydrophobic groups are present. Like, in case of substances with hydroxyl group, alcoholic solvents are used; for carbonyl groups containing compounds, acetone or ether is used, while for nonpolar compounds, toluene or hexane is used. In order to produce eluent of different polarities, the combination of polar and nonpolar solvents in varying ratios should be used.

2.5 Size Exclusion Chromatography (SEC)

This technique separates or fractionates a sample into different fractions on the basis of their size and molecular weight. For nonaqueous solutions, this approach may be extended to different polymers which is also called gel permeation chromatography (GPC). This may also be used in aqueous systems for distinguishing biomolecules. Instead it is then referred to as gel filtration chromatography.

2.5.1 Principle

A porous substance such as a polymeric gel or agarose beads with a diameter of generally 10 to 40 m is required for the chromatographic column. When the size of the pores is equivalent to the size of the molecules passing through, separation occurs; large molecules cannot pass through the pores (Fig. 3). SEC column or SEC is a process used to separate out different proteins and thus purify your protein from other contaminant proteins. The principle on which this works is that larger proteins will have a larger

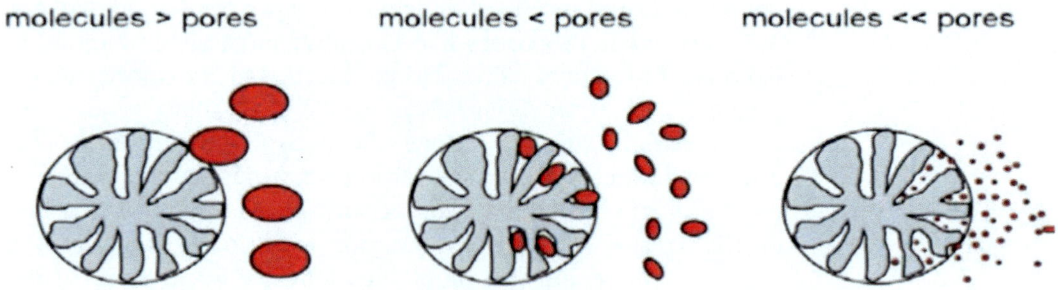

Fig. 3 Retention of molecules in size exclusion chromatography. (*Source: Andreas Manz and Nicole Pamme* [13])

hydrodynamic radius which will migrate differently through the gel matrix compared to smaller proteins which are much more globular in shape. So, basically this matrix is composed of beads which have very tiny pores of varying sizes. Now if we have a mixture of three proteins of 300, 100, and 50 kDa. Once it starts passing through the matrix of the size exclusion chromatography column, the smallest protein of 50 kDa size will be able to travel through all the different pores that represent in the matrix. So, it will travel through all the small pores present in the beads and as a result it will actually take a lot of time to pass from this point to the end point of the column. However, the large proteins that are also present as contaminants, they will not be able to enter all the pores. Some pores which are very small will not physically allow the large proteins to enter. So, they cannot spend time traveling through those small pores maybe they will travel through the comparatively larger pores. As a result, they have small path to travel and hence they will elute out earlier than the smallest protein. So, basically on applying a mixture of proteins through the top of the column by the time they are passing out of the column, the largest protein will come out first followed by intermediate sized proteins and the smallest proteins which had much more liberty to travel through all the pores in the matrix will elute out at the very end. Nevertheless, molecular size is no longer distinguished. Therefore, after a long transit period, all these small molecules are eluted together. Differentiation and separation take place only in a certain range of molecular sizes, usually between 2 and 200 kDa molecular weights, but the use of more specialized gels may increase to 1000 kDa. The size range depends on the pore size and their distribution in the gel matrix. The samples are collected in the tubes at the bottom of the column as fractions of a certain volume. However, in certain circumstances, the eluted fractions will be analyzed to determine both the sample fractions (often proteins) and the amount of sample (protein) contained in the fraction. Numerous techniques are often utilized, including the following: (1) spectrophotometric analysis of the fractions; (2) SDS-PAGE analysis of the fractions; and (3) assaying the fractions for a specific enzyme activity.

In SEC, retention volumes (V_R) are often used instead of retention time (t_R). The total volume (V_t) of the column is equal to the sum of the gel matrix volume (Vg), the gel particle volume (V_i), and the gel grain volume (Vo) as follows:

$$V_t = V_o + V_g + V_i \tag{7}$$

V_o, that is, void volume is the amount of liquid that is completely exempt from gel grains and believed to elute compounds. V_i is the product of dry weight of the gel (a) and the water regained (W_r), for example, $V_i = a \times W_r$. The elution volume

(V_e) is the volume required to elute the compound from a column, that is,

$$V_e = V_o + K_d V_i \quad (8)$$

where K_d indicates the part of the internal volume accessible to a specific compound and is independent of the column geometry.

$$\text{i.e.} K_d = V_e - V_o / V_i \quad (9)$$

Substituting the value of V_i in the above equation

$$K_d = V_e - V_o / aW_r$$

If $K_d = 0$, then $V_e = V_o$, that is, the elution volume would be the void volume.

$$\text{If} \quad K_d = 1$$
$$V_e - V_o = V_i \quad (10)$$
$$\text{Or} \quad V_e = V_i + V_o$$

The value of K_d is between 0 (molecule entirely eluted) and 1 (molecule having complete gel accessibility). K_d must be larger than 1 in order for the component to be adsorbed on the gel. Between V_o and V_o + V_i, all analyte molecules are eluted. The relevant molecular weight may be determined by charting the elution volume against the molecular weight of different markers and comparing the test compound's elution volume with the standard graph.

2.5.2 Media

Sephadex is the most common gel material in which dextran is crosslinked to form a hydrophilic and insoluble bead which when put in water swell considerably to form an insoluble gel. A number of other materials like Sepharose (stable between pH 4.0 and 10.0 and temperature 0–30 °C), Sepharose CL (stable over pH 3.0 to 14.0 and temperature up to 70 °C), Sephacryl, an allyldextran polymer covalently crosslinked with *N,N'*methylenebisacrylamide, various types of biogel-P made from polyacrylamide have been developed to attain better and improved resolution.

2.5.3 Applications

1. For the isolation of hormones, enzymes, polysaccharides, nucleic acids, proteins, and peptides in polymer mixtures, gel filtration chromatography is an extremely gentle process since it requires no harsh pH nor ion strength environments.
2. One of the most often used applications is the separation of salts and small molecules from macromolecules. This method of desalting is far quicker than dialysis, making it especially helpful for desalting labile substances.
3. The hygroscopic characteristic of dry gels enables the concentration of diluted solutions of macromolecules with molecular

weights larger than the exclusion limit. It provides a significant benefit for isolating proteins that are rapidly denatured by temperature changes.

4. Determination of molecular weight can also be done using this technique. It is important to calibrate the column with samples of species with known molecular weight.

   These above applications of gel filtration should take into consideration the following parameters:

1. Dimensions of the column: The column should be selected in such a way that for a specified volume, its length should be greater than its inner diameter. The separation or resolution can be improved by increasing its length.
2. Flow rate: Column resolution improves at a reduced flow rate for large biomolecules. Since the sample is permitted to diffuse freely in solution throughout the procedure, the peak diameter increases, especially for smaller molecules, with increased retention time. However, too large a flow rate can contribute to asymmetrical peaks.
3. Mobile phase: Undesirable ionic interactions occurring between the gel matrix and the separated molecules can be eliminated by using the mobile phase with an ionic strength <100 mM. Such interactions are often termed as "tailing."
4. Volume of the sample: An optimal sample volume should be <2% of the total volume of the column.

2.6 Ion Exchange Chromatography (IEC)

This technique separates and purifies a sample on the basis of their total charge. It is ideal for nearly any charged molecule including large proteins, short nucleotides, and amino acids. It is considered as the first step involved in the purification of proteins.

2.6.1 Principle

IEC is focused on the mutual competition between charged sample molecules and salt ions for the stationary phase charged functional groups. Assume that the negatively charged molecules in the sample bind to the column via positively charged functional groups present on the surface of the column, while the neutral and positively charged molecules are eluted from the column.

Elution of the adsorbed components is accomplished by increasing the ionic strength of mobile phase. By increasing the salt content or changing the pH of the mobile phase, the negatively charged analytes are desorbed and gradually eluted. IECs' stationary phase is often referred to as gel. On the stationary phase, agarose or cellulose beads with covalently linked charged groups are attached.

The functional surface groups are positively charged in anion exchangers, while the cation exchangers have negative surface groups. Diethyl aminoethyl (DEAE) and carboxymethyl (CM) are widely used ion exchangers. pH of the mobile phase determines the separation ability of such ion exchangers like DEAE, an anionic exchanger will be deprotonated, therefore neutralized and lose its activity at high pH. CM and DEAE work well enough at pH values from four to eight where a variety of biomolecular applications have the highest significance. Proteins are ampholytes, containing both basic and acidic groups. The total charge of a protein is the sum of the individual charge of its amino acid components. Their net charge either positive or negative depends on the pH of the solvent. The isoelectric value, pI, is defined as the pH at which there is no net charge on the protein. When working at a pH near the pI, the protein adsorption to the stationary phase is minimal. But where the pH varies greatly from that of the protein pI, the protein is highly charged and interacts strongly with stationary phase. The protein must be positively charged to get adsorbed in a cation exchanger like CM. The mobile phase pH must therefore be lower than the pI of the protein.

The protein must be charged negatively, in order to get absorbed on an anion exchanger like DEAE. The pH of the mobile phase should therefore be modified to surpass the pI of the protein. Buffer concentrations for adsorption phases are kept relatively low, between 10 and 20 mM, so as to reduce competition with buffer ions for binding sites. Phosphate and acetate salts are widely used buffers. A steady rise in ion strength or a change in the pH of mobile phase is required for the gradual desorption of immobilized components. For instance, salt gradients like NaCl are typically used for cation and anion exchangers.

The elution of the proteins occurs as the concentration of the salt is raised from 0 to 1 M or even higher. The elution of proteins occurs when a competition occurs between the salt ions and the proteins for the binding sites. At a lower ionic strength the weakly charged proteins get eluted, while the higher charged ones get retained and elute as the salt concentration is raised. All the sample components retained on the gel can be eluted or desorbed over a certain ionic strength. Desorption can also be done by a transition in pH. This facilitates a reduction in the net charge of the proteins or neutralization of the functional groups of the ion exchanger resulting in the desorption of the analyte as the interaction between the exchanger and the component gets eliminated or diminished.

2.7 Affinity Chromatography

2.7.1 Introduction

Affinity chromatography is a method for purifying biomolecules using their chemical structure or biological function as a basis. The chemical to be purified is covalently bound to a ligand (binding substance) and immobilized on a chromatographic bed material (matrix). Purification using this approach is distinct from all other methods because it does not depend on variations in the biological properties of the molecules being purified but rather on very accurate biomolecular identification. By introducing a specific ligand

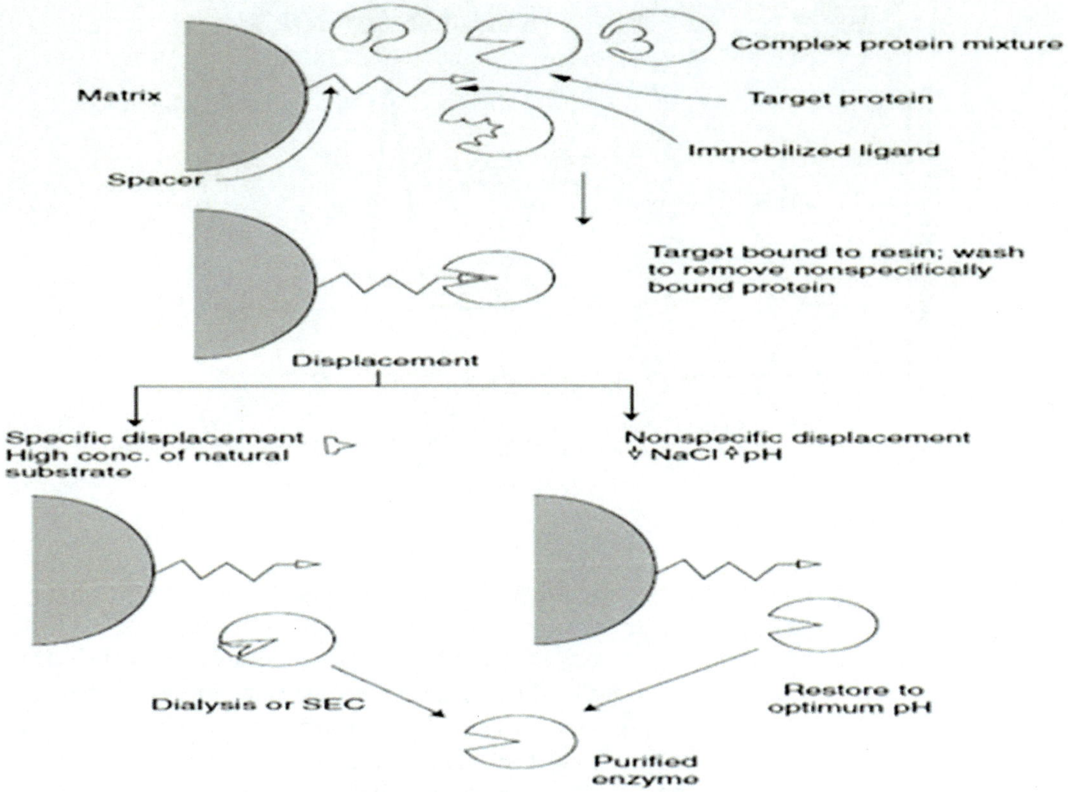

Fig. 4 Principle of affinity chromatography

such as an antigen to the stationary phase material, the matching antibody may be precisely and reversibly adsorbed. Not only does molecular identification exist between antigens and antibody but many other bonding partners include enzymes and co-enzymes, proteins from the receptor and hormones, or single oligonucleotide fragments and their matching counterparts. Affinity chromatography is an effective tool for purifying and isolating biomolecules even in low concentrations, with the best precision and selectivity of all chromatographic methods.

Principle

This procedure comprises sample introduction, adsorption, cleaning, and desorption (Figs. 4 and 5). Agarose or cellulose beads are covalently attached to the ligand molecules in the chromatographic column. The molecules that have a ligand affinity on the beads after the introduction of the sample are adsorbed and retained by stationary phase. All sample components with no ligand affinity are eluted from the column while subsequent washing allows the elimination of materials that are not explicitly bound. Finally, the adsorbed species get eluted from the column in the next step and is achieved by rupturing the non-covalent interaction acting between the biomolecules and the ligand. Several ways are possible,

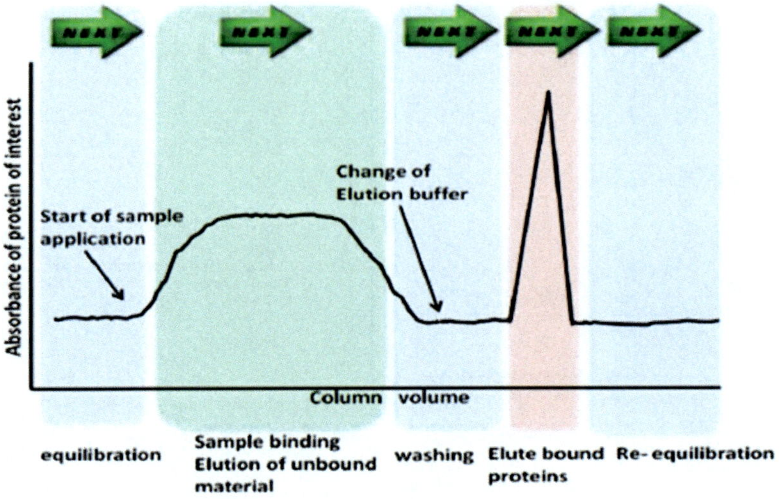

Fig. 5 Sequence of affinity chromatography

including lowering the pH, increasing ionic activity, adding a denaturing agent such as urea, or adding organic solvents (Fig. 4). This desorption process is nonspecific, since it elutes every bonded molecule identically. The presence of a species in the stationary phase that binds to the analyte more strongly than the ligand results in a particular desorption. The free ligand competes for protein binding sites on the stationary surface with the bonding ligand. Once attached to the free ligand, the protein is eluted from the column (Fig. 4). When the protein binds to the free ligand, it is ejected from the frame. After that, the separation matrix will be recreated. Affinity chromatography ligands may be classed as monospecific or group-specific. The former type of ligand has affinity for only one analyte. These ligands are synthesized and bonded to the stationary matrix material covalently. For instance, a specific hormone binds to its own binding receptor only. Affinity chromatography is also the best way to efficiently separate small amounts of biomolecules, while the latter type of ligand binds with the related proteins belonging to the same family of proteins. Immobilized lectins can bind conjugated proteins like glycolipids, glycoproteins, and polysaccharides, for example. Another example is the immobilized protein A which binds with the Fc region of an antibody. This Fc region is universally present in all the antibodies. These types of ligands can be commercially obtained with a wide range.

Advantages of Affinity Chromatography

1. Simplicity: No sophisticated and expensive chromatographic or electrophoretic apparatus are required.
2. Speed: The fractionation is usually rapid, saving time and preserving labile molecules.

Table 1
Different types of ligands and their specificity

Ligand	Specificity city
AMP	Enzymes with NAD cofactors and ATP dependent kinases
Arginine	Proteases such as prothrombin, kallikrein, clostripain
Cibacron blue dye	Dehydrogenase, kinase, serum albumin, preablumin
Heparin	Lipoproteins, growth factors, cytokines, coagulation factors (DNA, RNA)
Protein A	Fc region of immunoglobulins
Calmodulin	Calmodulin regulated kinases, cyclases, and phosphatases
EGTA-copper	Proteins with poly-histidine tails
Lysine	Plasminogen (r-RNA, ds DNA)
Lectin	Glycoproteins

3. Yield: Can be better than 90% under the right conditions.
4. High purification by functional selection: Can often obtain better than 90% purity from starting material comprising only 1–5% of the required molecule in one step.
5. Useful even to minor components: The procedure can also concentrate the minor components.

2.7.2 Types of Ligands

Table 1 shows the different types of ligands along with the specific compounds they adhere to.

2.7.3 Materials of Affinity Chromatography

Spacer arm: The spacer arm enhances the binding of the target molecule to the ligand by minimizing or eliminating the effects of steric hindrance.
Ligand: It is a type of molecule which can bind to a specific target or a group of target molecule.
Matrix: It is for ligand management and should be inert in nature both physically and chemically. The properties of a matrix should be as follows:

- Minimum nonspecific adsorption, for a successful affinity chromatography, the nonspecific interactions should be negligible.
- Derivatization of the hydroxyl groups of the sugar residues is important to ensure covalent binding of the ligand with the affinity media.
- The matrix should have a large open pore structure; this ensures that large molecules will attach more effectively. Generally, the inside of the matrix is accessible for ligand binding.
- In order to achieve rapid separation, it should have better flow properties.

Table 2
Various matrices used in affinity chromatography

Type	Examples
Inorganic materials/organic copolymers	Silica/hydrophilic copolymer
Bio-polymers/synthetic copolymers	Agarose-polyacrylamide
Inorganic materials	Porous glass, iron oxide
Synthetic copolymers	Polyacrylamide, polystyrene
Polysaccharides	Cellulose, starch
Biopolymers	Agarose

- It should be stable under different experimental conditions like low or high pH, presence of detergents and dissociating agents.

2.7.4 Various Matrices Used in Affinity Chromatography

Table 2 depicts different types of matrices that can be used in affinity chromatography.

Ligand

A molecule that can bind to a specific molecule or group of molecules reversibly, in order to ensure their purification by affinity chromatography. Two major factors play a substantial role in selection of a ligand: it should be having a high specificity for the target molecule along with binding it reversibly and it should have chemically modified groups capable of binding to the matrix without altering its binding capacity. In a free solution, the value of the dissociation constant (k_D) for the ligand–target complex should be ranging from 10^{-4} to 10^{-8} M. In case, when the dissociation constant falls outside the ideal range, alteration in the elution method can help in achieving the affinity chromatography successfully. In a solution, the ligand binds only to the target or desired molecule through a temporary bond. The formed ligand/molecule complex can be dissociated by altering the pH of the system.

Characteristics of Ligand

Ligand should form a covalent bond with the matrix such that its affinity for the protein does not get affected. The strength of the bond formed between the matrix and the ligand should neither be too strong nor too weak. The elution can be achieved by passing a buffer of low pH or high ionic strength from the column after binding.

Spacer arm

It is inserted in between the affinity ligand and the support matrix that facilitates and ensures its proper binding with the target molecule. The spacer arm (Fig. 6) is helpful in providing a more effective binding environment. A spacer arm is critical because in many cases, binding sites of the target molecules are buried in the structure and thus are restricted for ligand access or when the target molecule's receptor area is too large for the ligand. Thus, addition of a spacer

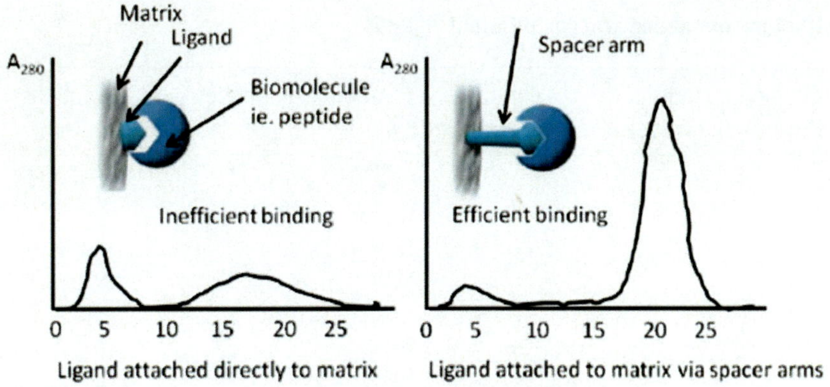

Fig. 6 (**a**) Direct attachment of ligand to matrix (**b**) Indirect attachment of ligand to matrix via spacer arms [14]

arm helps the affinity ligand to move away from the surface of the matrix, which allows a better and efficient binding of the target molecule. The length of a spacer arm (Table 3) is critical as improper size may cause binding failure or in some cases a nonspecific binding.

2.7.5 Immobilization of Affinity Ligands

- Prior to immobilization of ligand(s), the matrix is first activated by an easily reactive compound that is capable of binding with the ligand functional groups.
- The common reactive compound chemistries are primary amines, carboxylic acids, and aldehydes.
- Then, activated complex generates a covalent connection between the support matrix and the affinity ligand, resulting in the ligand immobilization.

Immunization of the affinity ligands can be classified as covalent immobilization and adsorption.

Covalent Immobilization

The hydroxyl groups present on the surface of Sepharose react with cyanogen bromide to form reactive cyanate ester groups. The primary amino groups present in the proteins react with this cyanate ester bond resulting in the formation of isourea bond (Fig. 7). This cyanate activated agarose acts as alternative to NHS-activated agarose. Cyanogen bromide should be handled in safety hood as it is a hazardous chemical.

Adsorption

Attachment of the ligand is site-specific and common examples are:

Specific Adsorption

- Adsorption of biotin containing ligands can be done by using streptavidin or avidin.
- Protein A/G is usually used for immobilization of antibodies.
- In specific adsorption method, the primary ligands, such as streptavidin, avidin, protein A/G are first immobilized on the solid matrix using covalent immobilization (e.g. amine-reactive method).

Table 3
Length and structure of spacer arm for different ligands

Chemical Groups on ligands (Protein, peptide, amino acids)	Length of spacer arm	Structure of spacer arm	Product
Amino	10-atom		HI Trap NHS-activated HP
	None	—	NHS-activated sepharose 4Fast Flow
			CNBr-activated sepharose 4B
			CNBr-activated sepharose 4 Fast Flow
	10-atom		ECH Sepharose 4B
Carboxyl	11-atom		EAH Sepharose 4B
Thiol	4-atom		Thiopropyl sepharose 6B
	10-atom	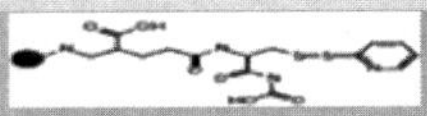	Activated thiol sepharose 4B
	12-atom		Epoxy-activated sepharose 6B
Sugar			
Hydroxyl	12-atom	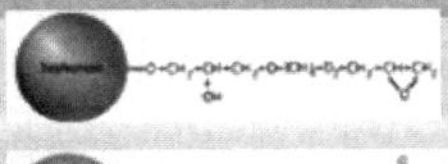	Epoxy-activated sepharose 6B
Amino	10-atom		HI Trap NHS-activated HP
	10-atom		ECH Sepharose 4B
	12-atom	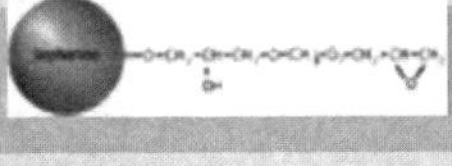	Epoxy-activated sepharose 6B
Carboxyl	11-atom		EAH Sepharose 4B
Polynucleotides			
Amino	None		CNBr-activated sepharose 4B
			CNBr-activated sepharose 4 Fast Flow
Mercurated base	4-atom		Thiopropyl sepharose 6B
Coenzyme, cofactors, antibiotics, steroids			
Amino, Carboxyl, Thiol or Hydroxyl	—	—	Use matrix with spacer arm (see above)

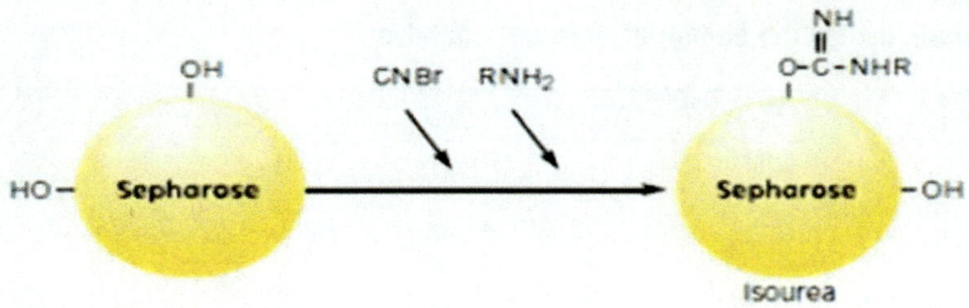

Fig. 7 Covalent immobilization

- Biotin or an antibody, as per the need, is then attached to them through adsorption.

Elution of Analyte

- Nonspecific adsorption.
- Various types of interactions occur between the ligand and the protein like non-covalent, hydrophobic, Van der Walls, hydrogen bond, and ionic bond.
- A number of treatments can be applied to disrupt these non-covalent interactions.
- The main objective of elution is to ensure that the protein is recovered in its most viable biologically active form. Thus, buffers of extreme pH or hydrophobic solvents should be avoided for elution purposes.
- Making slight changes in the composition of the elution buffer can serve this purpose like addition of 6 M urea can disrupt hydrogen bonds between the target molecule and the ligand, altering the pH of the elution buffer or incorporation of 0.5–1.0 M NaCl in the elution buffer disrupts the ionic interaction.

Elution of analyte can be accomplished by changing the composition of the buffer which facilitates the elution of the substance without affecting or harming the ligand. Another way is the use of high concentration of chaotropic agents or using extreme pH but it may lead to permanent or temporary damage to the substance or the ligand, while the last method of elution is called as specific elution. This is accomplished by the inclusion of a substance that competes for the binding site. This technique may be used to increase the specificity of media that include group-specific ligands. No specific sites are involved and the ligand molecules are adsorbed on the matrix surface by hydrogen bonding and coulombic or hydrophobic forces.

Conditions for the Elution of the Analytes

Table 4 depicts the conditions required for the elution of different analytes.

Table 4
Conditions for the elution of different analytes

pH	100 mM glycine–HCI, pH 2.5–3
	100 mM citric acid, pH 3.0
	50–100 mM triethylamine or triethanolamine, pH 11.5
	150 mM ammonium hydroxide, pH 10.5
Ionic strength and/or Chaotropic effects	3.5–4.0 M magnesium chloride, pH 7.0 in 10 mM Tris
	5 M lithium chloride in 10 mM phosphate buffer, pH 7.2
	2.5 M sodium iodide, pH 7.5
	0.2–3.0 M sodium thiocyanate
Denaturing	2–6 M guanidine–HCl
	2–8 M urea
	1% deoxycholate
	1% SDS
Organic	10% dioxane
	50% ethylene giycol, pH 8–11.5 (also chaotropic)
Specific competitor	>0.1 M counter ligand or analog (e.g. using gluthathione to elute GST-tagged proteins from immobilized glutathione agarose resin)

2.7.6 Modes of Chromatography

Lectin Affinity Chromatography

This technique is used with the aim of studying the glycosylation of a protein which occurs as a result of post-translational modification. Lectins are carbohydrate-binding proteins that have two or more binding sites for carbohydrate. They are categorized into five categories based on their monosaccharide specificity.

They have the strongest affection towards:

- *N*-Acetylneuraminic acid, fucose, *N*-Acetylglucosamine, galactose/*N*-Acetylgalactosamine, and mannose.
- The sugar residue present in the protein molecule binds with the immobilized lectin.
- The unbound contaminants are separated from the bound proteins by washing which ensures the elution of the purified protein.
- Lectins are a type of proteins that have affinity for sugars and are produced by molds, plants, and animals. Lectins are generally carbohydrate specific, that is, they recognize a specific kind of carbohydrate. They are mostly present in tetrameric form. Their subunit may or may not be identical and each identical subunit recognizes only a particular carbohydrate. As pea and soya bean are found to be abundant with lectins, hence lectins of plant origin are commonly used. These can be immobilized on CNBr-

Table 5
Common lectins used to fractionate glycoproteins

Lectin	Specificity	Useful element	Uses
Concanavalin A, from Canavalia ensiformis (jack bean) (Requires Mn^{2+} and Ca^{2+} for binding	α-D-Mannoside with free hydroxyl groups at C_3, C_4 and C_6	(a) 0.01-0.5 M methyl α-D-Mannoside (b) D-Mannose (c) D-Glucose	Separation of glycoproteins binding efficiency is reduced in the presence of a detergent
Lens culinaris	Terminal α-D-Mannoside or α-D-Glucoside	(a) Methyl-α-D-Glucoside (b) 0.15 M methyl α-D-Mannoside 0.1 M borate pH 6.5	Glycoproteins bind less strongly than Concanavalin. A binding take place in the presence of 0.1% (w/v) sodium deoxycholate. Useful for the isolation of membrane glycoproteins
Tritium Vulgaris	*N*-Acetyl-D-Glucosamine	0.1 M *N*-Acetyl-D-Glucosamine	A binding take place in the presence of 0.1% (w/v) sodium deoxycholate. Purification of RNA polymerase transcription factors
Ricinus communis	α-D-Galactoside	0.15 M D-Galactose	
Jacalin from Art ocarpus integrifolia	α-D-Galactoside	(a) O-linked sugar 20 mM α-methylgaiactoside (b) 800 mM D-Galactoside	(a) Separation of O-linked sugars (b) IgA from IgG
Galanthus nivalis	Multiple α(1–3) man nose residues	0.5 M Methyl-α-D-mannoside	Mouse IgM
Mannan binding protein. Requires Ca^{2+} (part of mammalian collectin family of lectins)	Terminal non-reducing sugar residues mannose, *N*-Acetyl-D-Glucosamine GlcNAc, fucose and glucose	(a) 25 mM mannose (b) *N*-Acetyl-D-Glucosamine	Mouse IgM

activated or NHS-activated agarose. Lectin and the glycoprotein interact with each other through nonionic bonds, so binding of glycoprotein to lectin column can be achieved without removal of salt after salt fractionation. The bounded glycoprotein can be eluted by altering the pH, reduction of the hydrophobic interaction by addition of ethylene glycol, use of borate buffer or carbohydrate in the elution buffer. Table 5 presents the common type of lectins that are used to fractionate glycoproteins along with their specificity, target elements and applications.

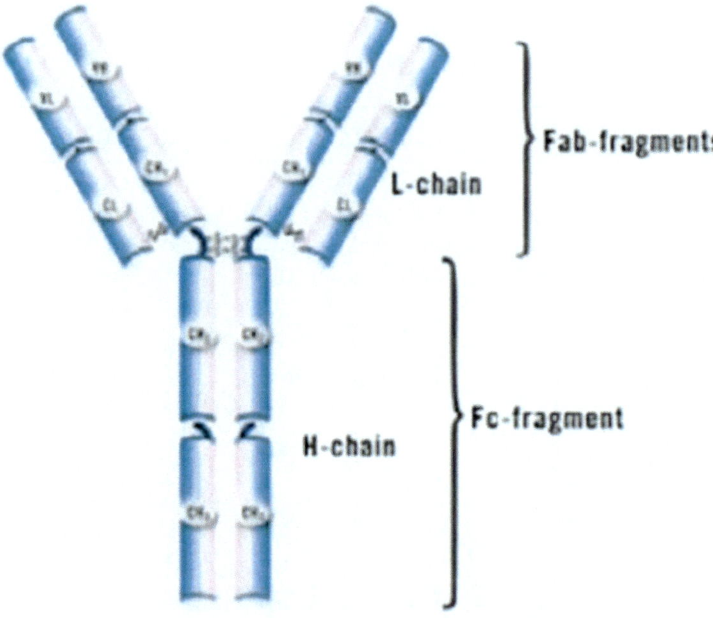

Fig. 8 Structure of immunoglobulin

Immunoaffinity Chromatography

Whenever a foreign molecule enters the body or the blood circulation, a series of reactions get triggered which results in release of certain proteins which are specific and bind to that foreign molecule. The synthesized protein is termed as an antibody, while the foreign molecule is termed as an antigen or immunogen. The interaction between antigen and antibody is highly specific and this forms the basis of this immunoaffinity chromatography. The antibodies synthesized by the animal against the antigen are generally polyclonal in nature, while monoclonal antibodies can also be prepared with the latter being more advantageous than the former ones. Primary amino group of antibody or protein antigen can be coupled to *N*-hydroxy succinamide (NHS) activated or cyanogen bromide (CNBr) activated-Sepharose. In tris buffer or phosphate buffer saline, antigen and antibody interact noncovalently (pH 7.2–7.4). Elution may be accomplished with either a low or a high pH buffer; the eluted fraction should be neutralized quickly to preserve the natural form of the protein (Fig. 8).

Metal Chelate Chromatography

Divalent metallic ions like cadmium, copper, mercury, zinc, or the transition metals ions like nickel, cobalt, manganese, nickel can be attached to iminodiacetate or tris (carboxymethyl) ethylenediamine substituted agarose (Fig. 9). The metal ions are capable of forming dative bond with certain groups of amino acids like tryptophan (indole group), histidine (imidazole group), and cysteine (thiol group). Higher the number of dative bonds greater will be the interaction and retention of the molecule or protein. The unbound

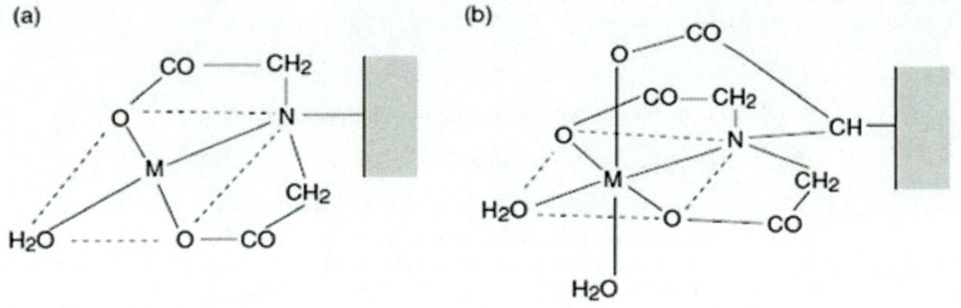

Fig. 9 (**a**) Tridentate iminodiacetic acid, (**b**) Quadridentate nitrilotriacetic acid

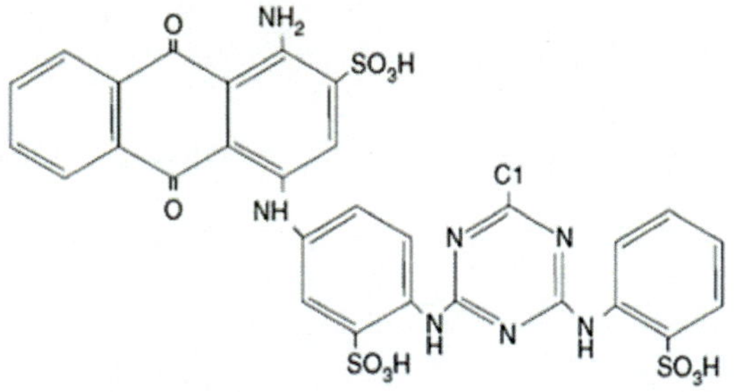

Fig. 10 Structure of Cibacron blue F3GA

or poorly bound proteins can be removed by washing, while elution of bound proteins can be achieved by either lowering the pH of the elution buffer or by incorporation of the chelating agents in it.

Psuedoaffinity Chromatography

Figure 10 shows the schematic of psuedoaffinity chromatography. Dye adsorbents such as Cibacron blue F3GA (Fig. 11) can be used as affinity adsorbents even though the dye bears no obvious resemblance to the true ligands. Therefore, it is referred to as psuedoaffinity ligands. Proteins that bind the dye are generally enzymes that use nucleotide cofactors such as ATP, NAD^+, coenzyme A. These molecules have in common partial structure and it is likely that sites on the enzymes that bind these structures are the sites that recognize triazine dyes.

Cibacron blue F3G-A is a triazine dye with ionic groups and conjugated rings. The dye binds to some proteins through hydrophobic and ionic interactions. Until recently, the kind of protein that binds to the dye and other proteins has not been established. Dye is inexpensive, adaptable to matrices, and stable. As it is not possible to ascertain that which protein will bind to dye, interaction between the protein and the dye can only be ascertained with trial-and-error methods. The binding of protein to the dye is done at a neutral to slightly alkaline pH (7.0 to 8.5). The dye can be immobilized on Sepharose 4B. Salt gradient is used for eluting the bound protein.

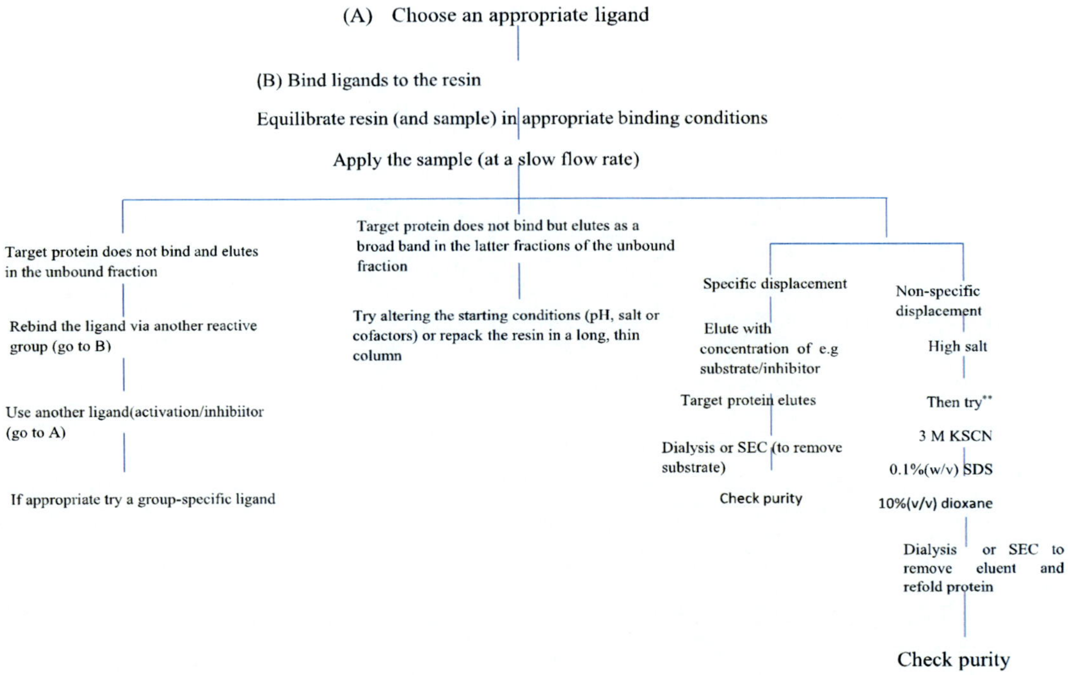

Fig. 11 Schematic of psuedoaffinity chromatography

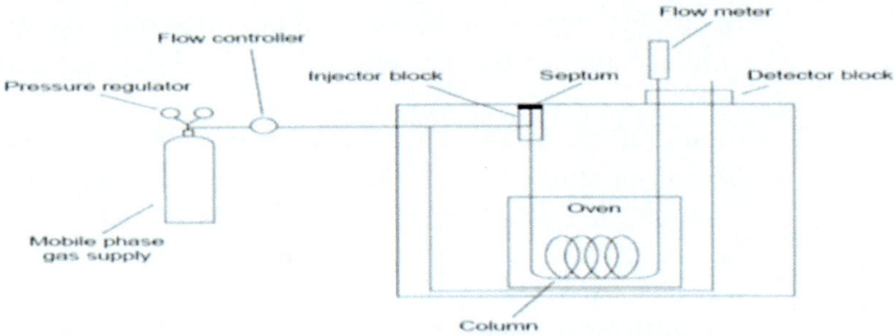

Fig. 12 Schematic diagram for a typical gas chromatography instrument. [6]

2.8 Gas Chromatography

This technique is widely used at both the industrial and academic level for analysis of various samples. It also comprises a mobile and a stationary phase. The stationary phase is made up of a liquid support suspended in a column, while the mobile phase is made up of an inert gas. Figure 12 shows the schematic diagram for a typical gas chromatographic instrument.

2.8.1 GC Columns

A chromatographic column offers a place to maintain the stationary phase physically. The layout of the column also determines the quantity of sample to be handled, the effectiveness of the separation, the amount of analytes that can be easily separated, and the time it takes to separate. Generally two types of columns are used in gas chromatography, namely packed and capillary columns. The former type of columns are made of glass, stainless steel, Cu, or Al, generally 2–6 m long, having an inner diameter of 2–4 mm. The sample is partitioned or divided into the gaseous mobile phase and the stationary phase, thus forming the basis of gas–liquid chromatography (GLC). In order to avoid the adsorption of solute molecules to exposed packaging material which degrades the quality of the separation, surface silanols are deactivated before stationary phases are covered by silanization with dimethyldichlorosilane or washed with alcohol (typically methanol). Capillary or open tube columns are made of the protective polymer-coated fused silica. The length of the columns can be up to 100 m with an inner diameter of around 0.25–0.5 mm. Capillary columns are classified as support coated tubular columns (SCOT) and wall-coated open tubular (WCOT). WCOT columns have a thin layer of a regular 0.25 mm thick stationary phase, coated on the inner wall of the capillary. In SCOT, a thin layer of a rigid support like a diatomaceous earth, covered with a liquid stationary phase is applied to the inner wall of the capillary.

Stationary Phases

The type of the stationary phase determines the selectivity of the GC system. GLC elution order in particular is defined by the boiling point of the solute and, to a lesser extent, by the interaction of the solute with the stationary phase. Solutes with significant difference in boiling points can be distinguished easily. On the other side, only if the stationary phase interacts exclusively with one of the solutes, two solutes of identical boiling points can be differentiated. Nonpolar solutes are usually most readily differentiated by the nonpolar stationary phase, while the polar solutes are distinguished by polar stationary phase. The key criterion for choosing a stationary phase is the solutes should have a low volatility, exhibit good polarity, should be thermally stable and inert to chemicals. Bonded or crosslinked stationary phases of capillary columns give superior stability. Stationary bonded phases are attached to the silica surface of the capillary. The cross-linking, which is done in the capillary column after placing the stationary phase, binds separate polymer chains, providing greater stability. The thickness of the stationary phase is another important feature of a gas column. The quality of separation increases for thinner films.

2.8.2 Sampling Techniques in GC

Sample Introduction

The application of the samples to the gas chromatograph is decided by three criterions. The sample or its component should be volatile in nature so that it passes through the column, the analytes of interest should be present in sufficient amount so as to be detected by the instrument and finally the sample should not weaken the separation on injecting. The column retains solutes with low volatility and continues to elute them when analyzing subsequent samples. The column condenses nonvolatile solutes which can degrade the column's efficiency. So, the nonvolatile analytes must be derivatized into a volatile component.

Solid and liquid samples must be in the gas phase before introduction into the GC.

Gas Sampling

Gaseous samples should be injected or transported into the GC column efficiently and carefully. Techniques like valves, gas tight syringes, gas tight bags, and canisters can be used for gas sampling.

Liquid Sampling

Liquid sampling is done using a syringe or by automatic liquid sampler. The column should not be allowed to get overloaded when injecting liquids with high expansion coefficients.

Solid Sampling

Solid samples should be dissolved in a suitable volatile liquid and then injected using a syringe. In these types of analysis, it is usual to drive the analyte from the solid by first increasing the surface area of the solid (perhaps by micronization) followed by strong heating. The evolved gases are then either sampled from the headspace around the sample, or all evolved gases are swept into the GC column.

2.8.3 Selection of Columns

Stationary Phase/Analyte Interactions

Dispersive: Also known as London forces/instantaneous dipole-induced dipole interactions. These types of interaction occur everywhere and among all the molecules.

Dipole–Dipole interaction: Such types of interactions are very strong and occur between permanent dipoles.

Dipole-induced dipole: Occurs between a dipole and induced dipole in the neighboring molecule.

Hydrogen Bonding: Interaction between molecules containing O–H, F–H, and N–H bonds and an electron pair rich molecule is termed as hydrogen bonding. This type of bond is very strong.

Stationary Phase Selection

1. Critical phase and temperature directly affect selectivity.
2. Principle of like dissolves like holds well.
3. Separates nonpolar analytes by using a nonpolar phase and vice versa.
4. Knowledge about what level or degree of polarity is sufficient enough to obtain satisfactory resolution or separation along with to prevent long retention time.
5. Separating compounds of mixed or intermediate polarity and functionality requires knowledge of the retentivity and selectivity of each phase.
6. May require fine tuning of the phase chemistry using the monomeric ratios.

Column Dimensions

As we know, that the resolution is given by.

$$R_s = 1/4\sqrt{N}\left(\frac{a-1}{a}\right)\left(\frac{k'}{k'+1}\right)$$

Efficiency is directly proportional to the internal diameter and length of the column.

Retention is directly related to temperature, film thickness, and internal diameter of the column.

Selectivity is directly related to temperature and phase.

Internal Diameter

- Has an effect on the column's efficiency, retention, carrier flow rate, capacity, and pressure drop.
- Inversely proportional to column efficiency—halve the diameter, double the efficiency, and gain a factor of 1.4 in resolution. Inversely proportional to analyte retention for isothermal but not for gradient.
- Consider gradient temperature programming in conjunction with pressure programming for constant flow.
- Column head pressure is proportional to the inverse square of the column radius.
- Column capacity rises with the internal diameter of the column but is also dependent on the phase type, film thickness, and nature of the analyte.

Effect of Internal Diameter

- Reducing diameter increases efficiency and therefore resolution. There occurs reduction in the column capacity.
- Increased efficiency will allow for the use of shorter column.
- Increase in the optimum linear velocity of the carrier gas with decreased column internal diameter giving shorter analysis time.

Film Thickness (d_f)

- Various factors like bleed, resolution, capacity, inertness, and retention are affected by d_f.
- A direct relation exists between the d_f and the retention time (1.5:1 for gradient).
- Thick stationary phase films provide retention for extremely volatile analytes.
- Increasing the film thickness results in volatile analytes being retained at or above ambient temperatures.
- The elution temperature can be increased by 20 °C, when d_f is doubled.
- Thinner film columns result in reducing the retention of late eluting analytes, while thicker film columns help in achieving a better resolution of early eluting analytes ($k < 2$).
- Increasing the d_f results in a lower resolution of analytes having k values ranging from 5 to 10.

- Films with a higher d_f have been found to bleed more.
- Thicker film columns reduce the peak fronting and have a higher analyte capacity.

Phase Ratio

- Ratio between column ID and film thickness.
- Use to keep retention time approximately constant while increasing efficiency (reduce column radius internal diameter) and reduce film thickness to keep ß constant, results in better and efficient separation as that of the original.
- Increase in the column radius or decrease in the film thickness results in decreased analyte concentration.

Selection of Column

- Capillary GC columns are commonly 10, 15, 20, 25, 30, 50, 60, or 120 m in length.
- Increasing the column length for the purpose of increasing resolution should be avoided.
- When the nature of the sample is unknown, column ranging from 25 to 30 m should be used. Generally, columns with shortest length should be selected to attain the required resolution. Another approach to improve the resolution is by altering the internal diameter rather than changing the stationary phase.
- The number of the analytes targeted for detection can be increased by reducing the internal diameter of the columns.
- The cost and time required to analyze using a column are proportionate to its length.
- Increase the film thickness when volatile analytes are present or lower the film thickness when strongly absorbed analytes are present.
- Using columns of smaller diameter without altering its phase ratio helps in improvement of the separation efficiency in the same timeframe.
- Increasing the column length also leads to the subsequent increase in the bleeding and head pressure of the column.
- Shorter columns (10–15 m) are better for samples containing lower concentration of the analytes.
- Columns of 50–60 m length should be utilized only when a significant number of components are required to be separated and only as a last option after the column internal diameter has been lowered, stationary phase is modified and temperature program has failed.

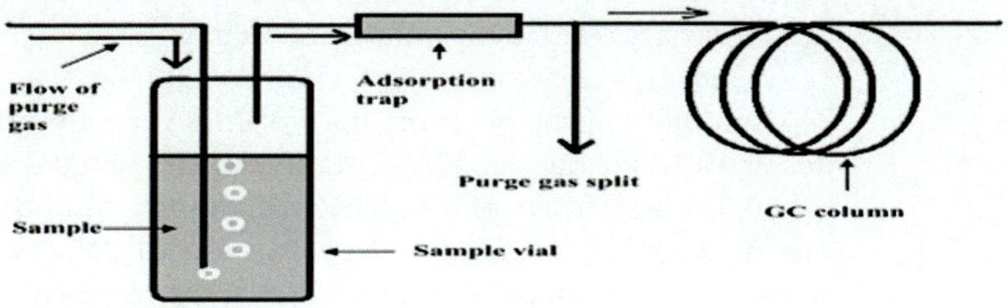

Fig. 13 Headspace structure [1]

2.8.4 GC Parameters

Please refer Subheading 1 for general chromatographic factors.

2.8.5 Head Space GC

Headspace GC (Fig. 13) is an excellent technique for qualitative and quantitative analysis and can be used for both liquid and solid matrices. It is used for analysis of dirty samples, that is, such samples which should not be injected into the instrument. It can be used for the trace analysis of the samples. Main problems associated with the headspace sampling involve the irreproducibility, lack of sensitivity, and decomposition.

Head Space Sample and Solvent Types

Liquids that can be analyzed on head space GC include insoluble components such as blood, paint adhesives, viscous or high boiling liquids. Nowadays capillary columns which are more resistant to the exposure of water are available; however, backflash is still a concern.

Solids: In case the solid is completely insoluble, headspace analysis may be the only option. For solids which are required to remain intact can be used in the pharmaceuticals.

Solvents must completely dissolve the analytes of interest. Water being nontoxic and easy to handle can be an excellent choice.

Headspace Analysis as an Equilibrium Technique

It is an equilibrium technique. Heating of the samples facilitates the release of the volatile analytes into the headspace. Not all analytes will evolve into the head space.

After equilibration,

$$K = C_L/C_G$$

$C_L =$ concentration of analyte in sample.

$C_G =$ concentration of analyte in the head space gas.

Compounds having low K values have a greater tendency to partition into the gas phase.

Headspace Sampling

Manual sampling of headspace gas is performed using a heated gas tight syringe, or an automated method is followed utilizing a heated gas loop and heater transfer line.

Equilibration

Analyte headspace concentration is dependent upon the ratio of the sample volume to headspace. To simplify the calculations, V_G and V_L are usually arranged to be equal. 10 mL of liquid sample in a 20 mL headspace vial is typical. It is critical to maintain strict control over the reproducibility of sample preparation, heating, agitation rate, and sample volume or weight ratio.

Analyte Solubility

The solubility of the polar organic volatiles is greatly affected as it decreases when the salt concentration in the sample is high, this results in the transfer of such analytes into the headspace. Salting out is generally accomplished by the use of chlorides, citrates, sulfate salts of sodium, and ammonium ions. For each component, the amount of the salting out effect varies. Compounds having lower *K* values experience minor alteration in the partition coefficient after addition of the salt. The effect of salt addition is greatest for polar volatile compounds in polar (aqueous) matrices.

Derivatization is a process which is carried out to enhance the volatility of a compound which subsequently results in the increase in the sensitivity and chromatographic performance. Presence of reactive hydrogen in acids, alcohols, and amines makes their analysis a bit difficult in the head space of GC. They can react with the analytical column or the surface of the injection port resulting in reduced response and tailing. Derivatization also reduces the chance of surface adsorption in the GC system. Some of the derivatization reactions include esterification, acetylation, silylation, and alkylation.

Standby: Sample is heated and agitated in small oven.

Pressurization: An inert gas is introduced into the vial headspace to increase the vapor pressure.

Loop filling/vent: Headspace gas flows through a gas sampling loop to vent.

Injection: Valves are altered and the carrier gas flows through the loop to sweep the sample through the transfer line and into the split/splitless inlet.

2.8.6 Detectors Used in GC

Detectors

The ideal detector has several desirable characteristics, including low detection level, linear response over a variety of solutes, tolerance to all solutes or selectivity to a single solute type, and flow rate or temperature insensitivity.

Features of GC detectors

Monitor a physical property of the analyte that must change for the detector to give the response.

Three important classifications

1. Concentration or mass flow-effect of carrier gas flow.

   Concentration sensitive (TCD, ECD) will have variable peak areas with changes in the flow rates.
2. Selective or universal-whether give response for a particular analyte or multiple analytes.
3. Destructive or nondestructive—allows further analysis—Multiple destructive detectors can be used if the eluent flow is split.

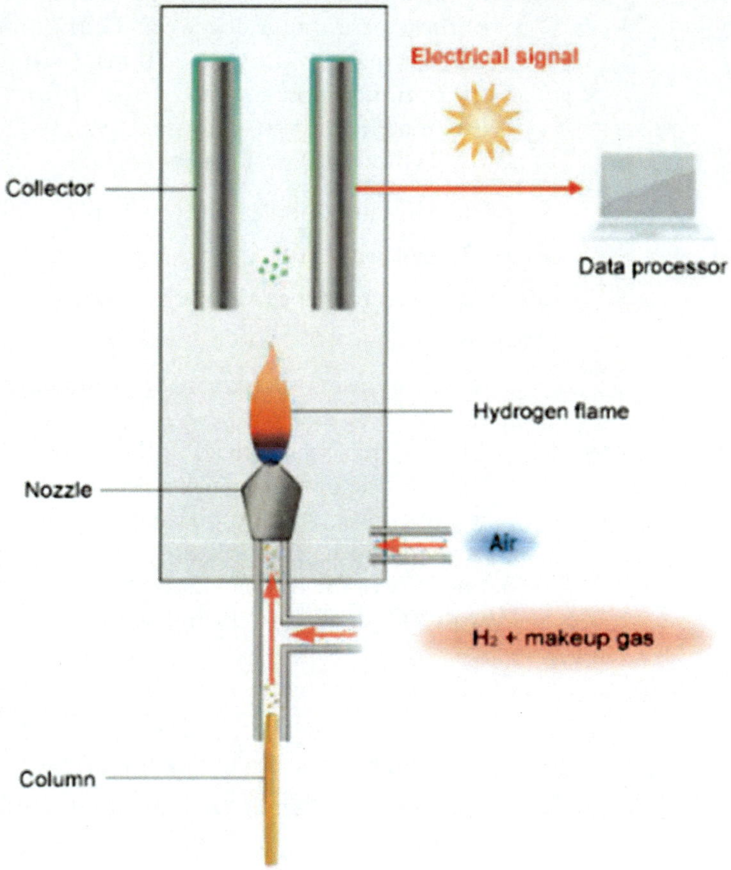

Fig. 14 Flame ionization detector (Source: Shimadzu)

Flame Ionisation Detector

Subjecting an organic compound into a mixture of H_2/air flame leads to the generation of a flame rich in ions and electrons. Applying a potential of approximately 300 V across the flame generates a small current. This current produces an analytical signal when amplified. This principle forms the basis of the **flame ionization detector (FID)** (Fig. 14).

Basics steps are as follows:-

1. The carrier and detector fuel gases get mixed at the tip of the anode producing a flame.
2. The potential difference (300 V) applied between the anode and collector should be electrically isolated.
3. The analyte gets ionized releasing cations, which travel towards the cathode generating a very small current of the order of pA.
4. Amplification of the generated signal is done and recorded as chromatographic peak.

Three types of gases are required in these types of detectors, viz. a fuel gas which is generally hydrogen, makeup gas generally helium or nitrogen and an oxidizer gas which is typically air. The flow rate of these gases ranges between 1 mL/min for carrier gas, 30 mL/min for makeup gas, and 300 mL/min for air. In order to ensure the optimum flow into the detector from the column, makeup gas is operated at 20–40 mL/min.

Typical FID performance

1. Minimum detectability: 50 ppb.
2. Linearity: 10^7.
3. Detector type: Mass, destructive.
4. LOD = 10^{-11} g C.
5. Response: organic carbon containing compounds.

Thermal Conductivity Detector (TCD)

The thermal conductivity of the mobile phase plays a critical role in the working of this detector. As the mobile phase exits the column, it passes over a tungsten–rhenium wire filament, whose resistance depends on its temperature, which depends on the thermal conductivity of the mobile phase. Helium is the mobile phase of choice with TCD owing to its higher thermal conductivity.

Features of TCD

1. Generates response to every analyte.
2. Two filaments in separate blocks with heated elements connected to a wheat stone bridge.
3. The pure carrier and the column effluent flow over separate cells.
4. The balanced Wheatstone bridge (Fig. 15) generates the baseline signal.
5. Rate of heat loss from the filament changes.
6. Current applied to balance the Wheatstone bridge which is recorded as a signal.
7. Sensitivity of the detector depends on the size of the cell.
8. Gases and inorganic compounds can be detected.

TCD Performance

1. Respond to all compounds.
2. 400 °C temperature limit.
3. Universal, constructive, nondestructive detector.
4. LOD = 10^{-9} g/mL, linearity of $<10^{-11}$ g C.
5. Susceptible to variations occurring in the carrier gas flow and the ambient temperature.

The Electron Capture Detector (ECD)

The **electron capture detector (ECD)** (Fig. 16) is a selective detector in which the emitted electrons ionize the mobile phase (nitrogen gas), leading to the generation of more electrons which lead to the generation of the electric current between the

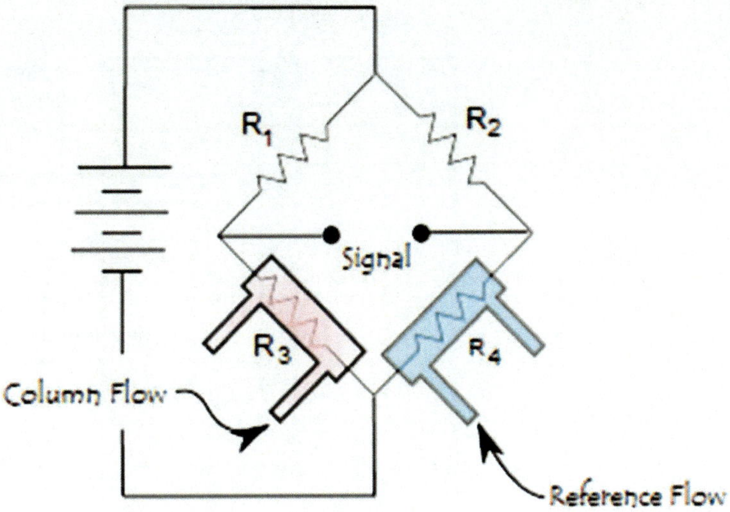

Fig. 15 Thermal Conductivity Detector [10]

electrodes. When a solute with a high cross section elutes from the column, it captures the electrons, leading to the reduction in the current. This decrease in the current produces the signal. ECD shows a higher selectivity towards solutes containing electronegative functional groups, such as nitro groups and halogens; however, it is insensitive to hydrocarbons, amines, and alcohols.

Features of ECD

- Measures electrical conductivity of the effluent after exposure to ionizing radiation.
- Sensitive to halogens.

ECD Operation

- The detector gases to be used should be free from moisture and be clean, water and oxygen being electronegative can generate noisy baselines.
- N_2 or Ar (5% methane) are used as makeup gases and their flow can be critical towards the sensitivity.
- Operating the detector in pulse mode improves the sensitivity and linearity.
- Constant current is maintained by applying the square wave pulse.
- Signal generated is proportional to the frequency.

Performance of ECD

- Highly selected for halogenated analytes.
- Temperature limit is 400 °C.
- As analyte elutes, the reduction in the current is compensated by the pulse frequency.

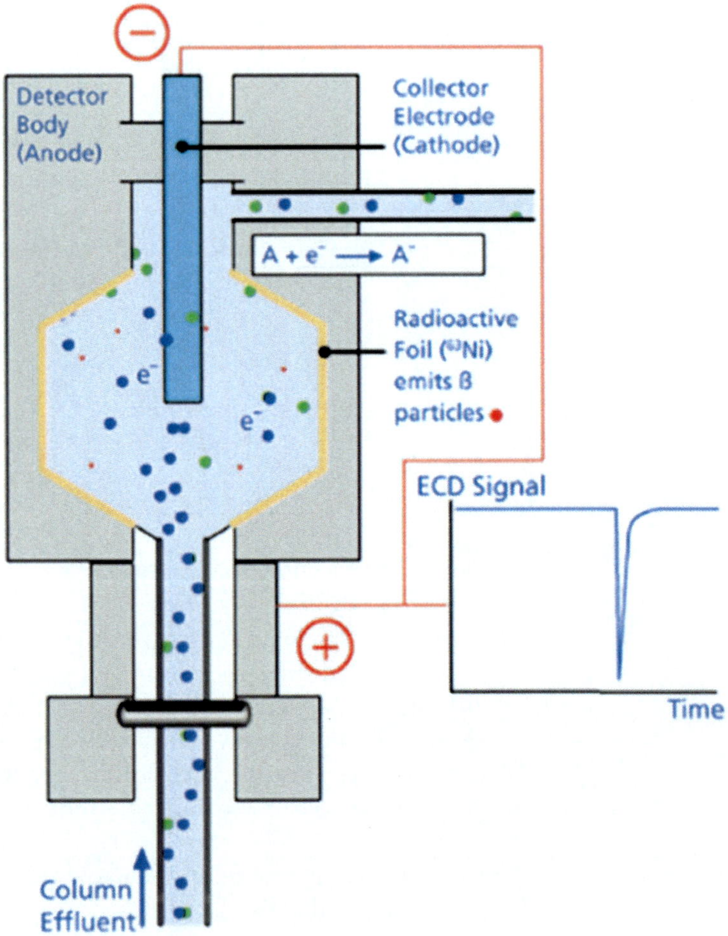

Fig. 16 Electron capture detector [11]

- Its performance gets highly affected by the leaks in the detector body and by the use of lower grade.
- Cleaning and reconditioning should be done cautiously.
- It is a selective type, concentration and nondestructive detector with LOD of 10^{-14} g/mL and a linearity of 10^7.

In the Flame Photometric Detector

In the **flame photometric detector** (Fig. 17) optical emission from phosphorus and sulfur when burned is measured. Photomultiplier tube then converts the light intensity into electrical signal.

Nitrogen Phosphorus Detector/Thermionic Detector

Compounds containing nitrogen or phosphorus elicit a response from the thermionic detector. Fourier transform infrared spectrophotometers (FT–IR) and mass spectrometers are two typical detectors of this category. It is also known as nitrogen phosphorus detector (Fig. 18).

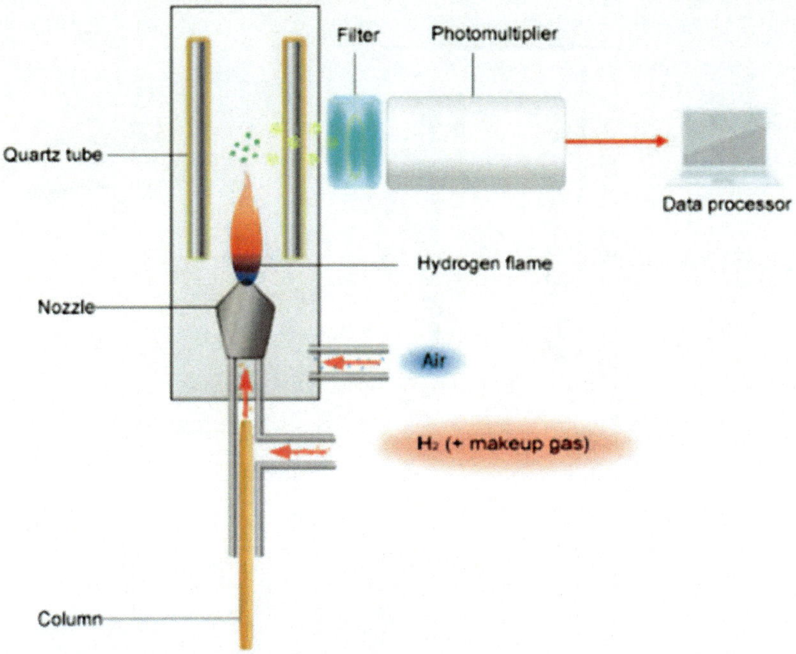

Fig. 17 Flame photometric detector (Source: Shimadzu)

Nitrogen Phosphorus Detector

- Similar to FID but with a different principle.
- Better sensitivity towards compounds containing phosphorous and nitrogen.
- This detector employs the use of rubidium silicate BEAD which emits thermionic electrons when heated.
- Hydrogen flow much lower than in FID detector—too low to sustain a flame at the jet trip.
- The plasma formed by the mixture of makeup gasses, carrier, oxidizer, and fuel gets ignited and burns partially around the bead heating element.
- The bead absorbs the partially combusted phosphorous and nitrogen containing compounds on its surface.
- This leads to the increase in the electron density at a given temperature or potential.
- This increase in the electron density results in the increase in the current, subsequently leading to a better and higher detector response.

NPD Operation and Optimization

1. Higher operating temperatures (260-350 °C) help in extending the life of the bead.

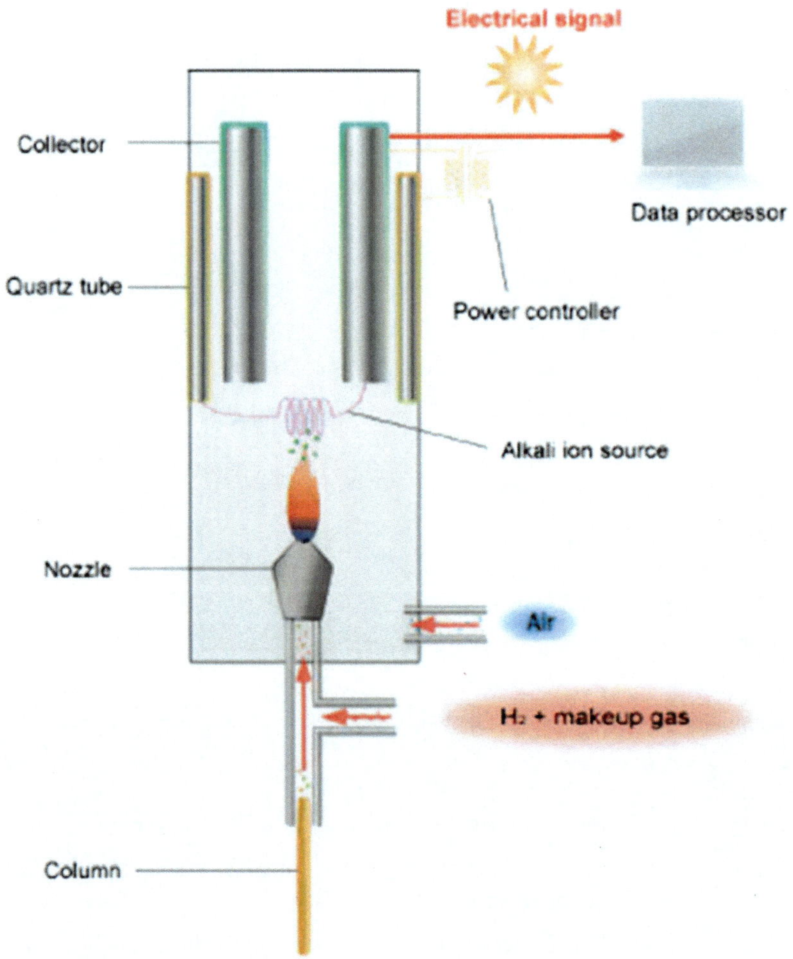

Fig. 18 NPD detector (Source: Shimadzu)

2. The bead power should be kept at the lowest at the start (2 V) which can be gradually increased up to 10 V if required.
3. Ensure that the flow of the hydrogen must be minimum so that it does not generate flame. In case of flame generated by hydrogen, the response from nitrogen will be zero.
4. The flow supply of hydrogen should be constant.
5. The performance of the detector decreases with time. Thus in order to obtain a suitable sensitivity a concomitant increase in the bead voltage is required.

NPD Performance

1. 500 times higher sensitivity than FID for N and P containing compounds.
2. 500 pg of pesticide easily achievable.
3. The carrier flow should be constant with a stable hydrogen supply.

4. 400 °C temperature limit.
5. Linearity 10^6.

2.8.7 Quality of Gas Used in GC

The carrier gas should be clean for column longevity and reduced baseline noise and good peak shape. Detector gasses may contain impurities which results in reduced sensitivity, a higher baseline noise, and an increase in the background signal. Air generators for GC are oil-free anyway. In-house air supplies may throw some oil into the stream. Also, a hydrocarbon filter is good insurance against traces of hydrocarbon from a less expensive or poor-quality tank.

Table 6 shows the recommended purifiers for various gas streams.

Gas Traps/Filters

There is a definite order for the installation of the gas traps. Hydrocarbon traps contain activated charcoal and remove contamination of oil mist from air generators. Oxygen (air) traps contain aluminum oxide and remove air from carrier and/or detector gasses. Molecular sieve traps contain molecular sieve and scavenge moisture. Most hydrocarbon traps have capacities of around 6–12 g (measured as n-butane) and will typically reduce the carrier hydrocarbon content to less than 100 ppb. Impure carrier may lead to reduced or poor sensitivity, generation of noisy baseline with a high background. There occurs bleeding of siloxane in the presence of oxygen/moisture by a typical back biting reaction which gets aggravated by increase in the temperature. Table 7 depicts the permitted levels of different impurities in carrier and detector gasses used in GC. Table 8 depicts the purity levels of different types of gases that are used in GC.

2.8.8 Split/Splitless Injection for GC

Split Injection

The proportion of the gas through the column and split line is termed as split ratio. It is the amount of the sample volume entering the column. Generally split ratios from 1:1 to 500:1 are employed during operation. For a split ratio of 100:1, for every hundred parts only one part enters the column and the remaining part is vented out. A higher split ratio means that the volume of the sample entering the column will be less. This avoids the overloading of the column, fronting of the peaks, poor area reproducibility and generates sharper peaks.

Effect of split flow on peak shape

1. With an increase in the split flow, increase in the liner flow (amalgamation of split and column flow) occurs.
2. Gaseous sample gets transferred rapidly into the column.
3. This results in the decrease in peak or analyte bandwidth at the column head.
4. The analyte band will disperse during elution but initial bandwidth has impact on peak efficiency.

Table 6
Recommended purifiers for various gas streams

Gas stream	Recommended buffer
Carrier gas	Hydrocarbon, moisture, O_2
FID, NPD, FPD (fuel gas)	Hydrocarbon
FID, NPD, FPD (oxidiser)	Hydrocarbon
ECD (reaction and makeup gas)	Moisture and O_2
TCD (make up)	Moisture and O_2

Table 7
Permitted levels of different types of impurity

Impurity	Certified gas levels (ppm) (99.9995%)	High purity gas levels (ppm) (99.995%)
Oxygen	1	3
Moisture	0.5	1.5
Hydrocarbons	0.5	1

Table 8
Purity levels of different types of gases [2]

Carrier Gases and capillary make up gasses	
He	99.9995%
N_2	
H_2	
Ar/CH_4	
Detector support gasses	
H_2	99.9995%
Air (dry)	Zero-grade or better

Split injection is used as the default vaporizing injector which is used for the non-trace analysis of volatile samples. The flow of the gas (particularly split flow) should be maintained carefully. Do not forget septum purge flow. The increase in split flow leads to the improvement in the peak shape, reduces the risk of thermal degradation as the analyte residence time gets reduced, reduces the analytical sensitivity, and reduces the column loading. Table 9 presents the advantages and disadvantages of split type of injection.

Table 9
Pros and cons of Split injection

Pros	Cons
Protect column from non-volatile sample components	Analytes susceptible to thermal degradation
Narrow analyte band on column	Liner geometry dictates injector settings
Rugged design	Suffers from discrimination
Simple to use	Not suitable for ultra-trace analysis
Easy to automate	

Table 10
Pros and cons of splitless injection

Pros	Cons
Simple to use	Conditions should be carefully optimized
Rugged design	Backflash can occur
Excellent for trace analysis	Thermal degradation of the analytes can occur
Lower risk of discrimination between analytes than split mode	
It can be easily automated	

Splitless Injection

When the flow of the split line is off or if the injector is in a splitless mode, the whole sample reaches the column. Vapors from the sample are trapped at the column's head (solvent and thermal effects). To commence elution, the column temperature is programmed. When the split line is switched on, the injector is emptied to the point where the analyte is transferred to the column. Table 10 presents the advantages and disadvantages of splitless injection.

Flow Through the Liner Column Flow During Splitless Phase

As the analyte can take minutes to get transferred to the column, it may lead to unacceptably wide peaks, so care should be taken to avoid it. This can be avoided by focusing the analytes onto the head of the column by the virtue of thermal and solvent focusing effects. Splitless time needs to be carefully considered. Initial temperature of the oven must be 150 °C lower than the analyte boiling point.

2.8.9 Temperature Programming in GC

Temperature Control

The temperature of the column plays a vital role in a successful gas chromatographic separation, so it should be controlled properly. This is achieved by a thermostat oven in which the column is placed. In an isothermal separation, the temperature of the column is constant which is determined by the nature of the solutes. The temperatures are usually marginally lower than for the lowest boiling solute to maximize the contact of the substances with the stationary phase. The temperature is steadily elevated as the separation occurs at a constant pace or through a sequence of phases. Temperature affects selectivity (α), retention (k), and efficiency (N) to a lesser extent. Clausius–Clapeyron equation relates analyte vapor pressure and temperature (inverse of temperature)

$$\ln\left(\frac{P_1}{P_2}\right) = \frac{\Delta H_{\text{vap}}}{R}\left(\frac{1}{T_1} - \frac{1}{T_2}\right)$$

Plot of log k vs. $1/T$ shows:

1. As temperature decreases retention increases.
2. Lines are not parallel; therefore, selectivity between analytes alters as a function of temperature.
3. As lines diverge at lower temperature it indicates that GC separations are better at lower T.

Isothermal and Temperature Programmed GC

As the name suggests, in isothermal conditions, the oven temperature remains constant during the analysis. This method helps in the analysis of the compounds that have similar retention characteristics and no significant difference exists between their boiling points. Meanwhile it may lead to certain problems like late eluting peaks, poor sensitivity, poor peak shape, and longer retention times.

Temperature programmed GC separations start with a low oven temperature which is increased during the analysis.

- All peaks elute with approximately the same width.
- Peak shape of the later eluting peaks is improved.

Temperature Programming in GC

An increase in oven temperature of 30 °C reduces analyte retention by half (for both isothermal and gradient). During a temperature program, the analyte migration through the column doubles for every 30 °C increase in temperature. Analytes pass through half of the column at 30 °C prior to elution and three-fourths of the column at 60 °C prior to elution. Every analyte takes the same amount of time to go through the column. Analytes are first stuck at the column's head before vaporizing and partitioning through the column. Isothermal mode of operation is possible, if the peaks elute over less than 25% of the temperature program time. As a rule, isothermal temperature is generally kept at 45 °C lower than that of the elution temperature of the final elute.

Initial Temperature and Hold Time

Resolution of the early eluting peaks is affected by the initial hold time and temperature. If the instrument is to be operated without cryogenic cooling, the oven temperature should not exceed 40 °C practically. The initial oven temperature can be reduced in place of addition of an initial hold for poor resolution of early eluting peaks. One precaution to be taken is that if one is working at an initial oven temperature of 30 °C less than the solvent boiling point, initial hold may be required to improve resolution of volatile analytes (30 s and increase if needed).

Optimum Ramp Rate

Ramp rate is the rate at which the temperature is increased during the temperature program. Mid eluting analytes, that is, eluting at the mid of the chromatogram, mostly get affected by the ramp rate. Optimum ramp rate is equal to 10 °C per void time. It can also be optimized at a rate of ±5 °C/min steps, if required. Insignificant or drastic changes occur if lower or higher steps are used, respectively. The overall analysis time includes sample preparation, sample introduction, separation and detection, cooling and equilibration, and reporting. Minimize the time taken to separate the peaks.

- Maximize the number of peaks which can be separated on a column.
- Increase carrier gas flow rate (F).
- Use a faster carrier gas (hydrogen-sweet spot for many hydrogen separations in the range 100–149 cm/s).
- Reduce column length (L).
- Reduce column diameter (d_c).
- Reduce stationary phase thickness.
- Increase temperature program heating rates.
- Optimization depends on analysis goals.

2.8.10 Selection of GC Column

If no information is available on which stationary phase to utilize, begin with a 100 percent methyl (PDMS) or a 5% phenyl phase. The stationary phase utilized should have a polarity comparable to that of the solutes. While this strategy has been proved to be more effective, it does not guarantee that we will always find the optimal stationary phase using this method. Utilize the least polar stationary phase possible while maintaining acceptable resolution and analysis times. If the dipole or hydrogen bonding strengths of the poorly separated solutes vary, switch to a stationary phase with a different quantity (not necessarily greater) of the dipole or hydrogen bonding interaction. Co-elutions that are not wanted may also occur when the stationary phase is changed, and hence the new stationary phase may not yield a higher overall resolution. Avoid stationary phases that have functionality that may generate a large response when used with a selective detector. A mixture of 100% methyl or 5% phenyl, 50% phenyl, 14% cyanopropylphenyl, and wax (PEG) covers the broadest range of selectivities with the fewest columns. When gaseous samples must be analyzed at temperatures greater than the ambient column temperature, PLOT columns are often employed. MS columns with little bleed have greater temperature tolerances and are often more inert. The lifetimes of nonpolar stationary phases are greater than those of polar stationary phases.

2.8.11 Derivatization in GC

Purpose of Derivatization in Gas Chromatography

Increase Volatility

Primarily via the reduction in intermolecular and ionic interactions between polar groups such as –OH, –COOH, –NH, –SH.

Improve Chemical/Thermal Stability

Through conversion of a reactive/labile functional group to one which is non-labile.

Improve Chromatographic Properties

Overcome peak tailing due to polar interactions between polar analyte functional groups and exposed silanol groups in the inlet or capillary column. Separation of derivatized chiral isomers using standard GC phases.

Improve Sensitivity/Selectivity in Quantitative MS Analysis

To produce derivatives with stable, more abundant parent ions or higher mass fragments to improve specificity and sensitivity-especially in GC-MS selected ion monitoring (SIM) experiments. Improve sensitivity/specificity in some common GC detectors such as ECD.

Increase Structural Information

To promote production of structurally informative fragments or label functional groups to track fragmentation pathways (halogenated derivatives).

It enables the detection of specific functional groups.

It facilitates the generation of a molecular ion.

Characteristics of Ideal Derivatization Procedure

1. To accomplish the desired modification.
2. To proceed quantitatively and reproducibly.
3. To distinguish and separate products from the starting materials.
4. It should not cause any rearrangements or structural alteration during the formation of derivative.
5. It should produce a derivative which is stable over time.

6. To be in line with the simple lab techniques which will be applicable to large number of analytes.
7. Should not involve any hazardous reagents and reactions.

Silylation

Reaction is not instantaneous and usually requires high temperatures of greater than 70 °C and time typically 30 min to 24 h.

Functional groups that are sterically hindered need more stringent conditions and reagents.

Typically, samples do not need extra sample cleaning.

Silylation is often used to derivatize –OH and –SH groups.

Acylation

The reaction replaces active hydrogen with an acyl group using a carboxyl or carboxylic acid derivative.

Typically used to detect MS fragmentation pathways or improve detectability of the compound.

Typically used with –OH, –SH and –NH compounds to produce the ester, thioester, or amide.

The reactions use very active chemicals, which may need further cleaning prior to injection.

Scavengers may be required to react with excess reagent.

For analysis, the products are stable.

Common application is the use of halogenated acylating reagents to improve detectability—especially with ECD detectors.

Alkylation

Replaces an active hydrogen (one not bonded to a carbon item) with an aliphatic or aliphatic/aromatic group.

Often used to convert alcohols to ethers, or carboxylic acids or esters to impart stability and improve chromatographic performance or detectability.

May also be used as the first step in multiple derivatization reactions.

Issues with Derivatization

- Formation of multiple products.
- Irreproducible yield.
- Need for multiple derivatization reactions.
- Possibility of degradation, rearrangement reactions, artifact formation concentration of impurities.
- Necessity for removal of excess reagents.
- Of special interest are derivatization reactions which can be carried out in situ or in an automated fashion which can help to improve reproducibility and offer unattended operation.

2.8.12 Fast Separations in Capillary GC

1. Accelerate the injection procedure.

   Advantages

   Simple to do with a high split flow.

   Drawbacks

   May result in the reduction of the analytical sensitivity/or lead to back flash.

2. Decrease column internal diameter.

   Advantages

   Shorter columns (no of plates per column higher), faster optimum carrier gas linear velocity.

   Drawbacks

   Lower injection volume, requires high inlet pressure and reduced column capacity.

3. Increase carrier gas velocity.

   Advantages

   Impressive speed gains employing vacuum-outlet techniques are easy to accomplish with increased head pressure and decreased interior column diameter.

   Disadvantages

   The instrument must be able to provide high head pressure, which is not yet completely proven technology for vacuum applications.

4. Decrease the length of the column.

   Advantages

   Standard internal diameter columns, low pressure drops, standard hardware.

   Disadvantages

   Reduced power of resolution.

5. Carry out isothermal separation.

   Advantages

   No cool down cycle, highest resolution power.

   Disadvantages

   For samples with a limited volatility range, no thermal or chemical focusing feasible. Late eluting compounds can cause column contamination.

6. Enhance rate of oven programming.

   Benefits

   Elutes the whole volatility range in the shortest possible time.

   Disadvantages

   Potential increases in resolution and selectivity may need additional hardware.

7. Use alternative carrier gas (H_2 > He > N_2).

   Advantages

   Same efficiency with shorter columns. Time and cost savings when using H_2 gas.

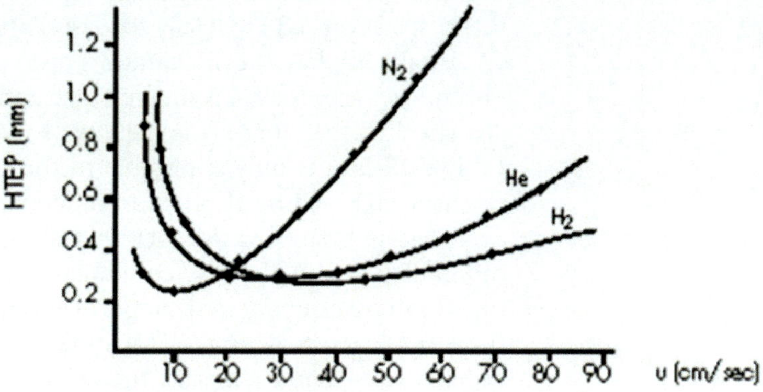

Fig. 19 van Deemter plot for N_2, He, and H_2 [2]

Disadvantages

There is a possibility of safety concerns with H_2, and the initial expense of acquiring a hydrogen generator is rather high.

8. Reduce the amount of dead space in the GC system.

Advantages

Optimize peak form and efficiency, which is simple to do with a smaller internal diameter liner and proper column placement in the GC inlet and detector.

Disadvantages

Purchase of a new intake liner and injection volume restriction to prevent "backflash" may be required.

2.8.13 Selection of Carrier Gas

Helium gas is expensive, has finite supply. MS and pulse discharge-based detectors (PDD, HID, BID) prefer helium because of the detection principle.

Carrier Gasses for GC

Hydrogen

Generated in situ in the lab. Gas is flammable, explosive, and reactive. There are compatibility issues with MS detection—high background.

Nitrogen

Slow but efficient, generated in situ in the lab.

As per the Van Demeter equation (Fig. 19):

$H = A + \frac{B}{u} + Cu$ where A represents eddy diffusion, B represents molecular diffusion, and C represents mass transfer. From the equation, it is evident that minimum H is required for the highest efficiency. From the figure it clear that the N_2 is having the minimum H at the lower carrier gas velocity. Hydrogen is the best carrier gas when highest linear velocity is required without sacrificing the efficiency. Shape of the curve is determined by the column internal diameter and gas type. Flatter the curve, easier it is to be operated at optimum linear velocity without compromising for the loss of separation. It is less critical when operating in constant flow or linear velocity mode. Hydrogen is best suited when higher linear velocity is required especially for faster analytes.

While N_2 has the maximum chromatographic efficiency, the optimal velocity for H_2 is much larger, resulting in fourfold quicker analytical times when H_2 is employed as the carrier gas. The term Eddy diffusion is only applicable to the packed GC columns as in capillary GC column this band broadening phenomenon does not occur. It is based on the fact that an analyte molecule can take different paths in the GC column. Owing to the inhomogeneity of column packaging and the particle size of the packing material, these many paths emerge. This multi-path effect appears to make the analytical band wider as they travel around the column, with different analytes taking shorter or longer paths. Eddy diffusion can be minimized by

- Pick columns that are well (tightly) packed.
- Use of smaller particles of stationary phase.
- Using particles with a limited size distribution.

It should be noted that the smaller particles produce higher column pressure.

Longitudinal Diffusion

Due to the concentration gradient at the band exterior margins, a band of analyte molecules embedded in the mobile gaseous phase appears to spread in every direction. The broadening of the analyte band is called longitudinal diffusion since the largest scope for expansion lies along the flow axis inside the tubes. In the sample inlet and detector, the band can also broaden, but the worst effects can be seen in the column. The length of the column defines the sample plug diffusion; the longer the sample remains in the column, the wider the column is and the greater the resultant peak. At low mobile phase velocity, the longitudinal diffusion has a much bigger effect. The consequences of this particular factor will therefore be minimized by the use of high linear velocity (high moving phase flow speeds with smaller internal diameter columns).

Longitudinal Diffusion Can Be Minimized by

- Ensuring proper installation of the column into the injector and the detector.
- Ensuring that the inlet liner and temperature are appropriate.
- Using a carrier gas having a low diffusion coefficient.
- It should be noted, however, that nitrogen is not the optimum carrier gas for capillary GC.
- Using shorter columns and increased flow rate for the mobile phase.

Stationary Phase Mass Transfer

The solute's mass transfer in either the mobile or stationary phase is represented by the C term in the van Deemter equation. Mass transfer explains primarily how easily the analyte molecules diffuses in the mobile or stationary phase. For example, fast solute sorption and desorption hold the solute molecules together and decrease band broadening.

Mass transfer in stationary phase can be reduced by:

- Ensure that all the analytes get eluted within a reasonable period ($k < 20$), make use of temperature programming.
- Using thin stationary films.

Mobile Phase Mass Transfer

Band broadening due to mass movement happens across the narrow GC column due to the nonturbulent fluid flow of the analyte band. The parabolic flow profile that arises as the analytical molecules moving ahead of the wall in the center of the column may result in an insufficient cross-channel combination in the gas phase. By using small columns of internal diameter, mobile phase mass transfer effects can be reduced.

2.9 High-Performance Liquid Chromatography (HPLC)

HPLC is a separation tool in which we can separate a mixture of complex analytes into its individual components. HPLC system includes the solvent handling unit, the degasser unit, the high-pressure pump, the sample injector, the column, column oven, and detector. The separation of the analyte depends on the nature and type of the phases used viz. the stationary phase and the mobile phase. Generally two modes are employed for separating the components, that is, separation by the normal phase and the other being separation by reverse phase. The nature of the stationary phase in normal-phase HPLC is generally silica based which are more polar in nature when compared to reverse-phase chromatography wherein the modified silica particles are used. In C18 stationary phase, the silica surface has been chemically modified to incorporate a C18 alkyl chain which makes the stationary surface more nonpolar or hydrophobic in nature. Today more than 95 percent of the separation of drug and drug-like molecules is carried out employing reverse-phase chromatography. In reverse-phase chromatography, the mobile phases consist of organic aqueous systems, the organic component of the mobile phase is generally methanol, acetonitrile, or a mixture of the two and the aqueous component of the mobile phase mainly consists of water. However as most of the drug molecules are weak acids and weak bases, their extent of ionization depends on the pH of the media. In such cases the water can be replaced by a suitable buffer. Typical buffers recommended in HPLC include the acetate, formate, citrate, and phosphate buffers. In addition to this, we can also incorporate a small percentage of about 1–2% of an organic modifier to modulate the polarity of the stationary phase. Typical modifiers used include isopropyl alcohol, tetrahydrofuran, and dimethylformamide and also a small concentration of an acid or a base or a salt may be also used. Typical acids used in HPLC include glacial acetic acid and orthophosphoric acid and in case of bases, triethylamine is preferred. In certain separations, an ion pairing reagent may also be incorporated. All solvents which are used should be of HPLC grade. All reagents and chemicals which are used for preparation of buffers should be of analytical or comparable grade. Water which is used should be of HPLC grade or triple distilled water. In case of HPLC, peak resins are used; therefore, strong acids like concentrated sulfuric acid, concentrated nitric acid, and strong organic solvents like chloroform, acetone, dichloromethane, and dimethyl sulfoxide should be avoided. Also, mobile phase containing halogens like sodium chloride, potassium chloride, and ammonium chloride should be avoided as they corrode the stainless-steel components which are used in HPLC. In HPLC, pretreatment of the mobile phase is very necessary. Pretreatment is carried out in two stages: degassing followed by filtration. For degassing, an ultrasonic bath is usually used. The mobile phase can be transferred to a suitable beaker and placed into the ultrasonic bath and ultrasonicated for a period of 3–5 min. Ultrasonication for longer periods is not recommended as this can lead to an increase in the temperature of the system and may affect the stability of the thermally unstable drugs. Ultrasonication is carried to degas the mobile phase; otherwise, this can result in the fluctuation of the pressure of the pump and can also affect the baseline of the detector. After ultrasonication, vacuum filtration is carried out. The vacuum filtration unit is loaded with a membrane filter. Filtration is carried out under vacuum and typically in HPLC, nylon and polyamide filters are preferred. Usually the 0.5 micron membrane filter is used for the mobile phase filtration and the 0.2 micron membrane filter is used for the sample or standard filtration. Once pretreatment of the mobile phase is carried out, they can be transferred to the solvent reservoirs which are mounted on the top of the HPLC system. HPLC systems are commercially available in two formats: the isocratic HPLC and the gradient HPLC system. In isocratic HPLC, during the entire separation process the composition of the mobile phase is constant, while in gradient HPLC, the composition of the mobile phase can be changed as a function of time. The inlet filters or the sanction filters are immersed into the mobile phase in the solvent reservoir. These inlet filters are made up of stainless steel of 10 micron porosity and help to entrap any particulate contaminants that may be introduced into the mobile phase when it is transferred to the solvent resource. The tubing which is used to connect the solvent reservoirs to the high-pressure pump is made up of Teflon or polytetrafluoroethylene. The HPLC system also comes with an inbuilt degassing unit which helps to further remove any air bubbles that may get entrapped in the system. The purpose of pump is to deliver the mobile phase to the column under very high pressure. It also helps to give constant and non-versatile delivery of the mobile phase to the column. The purge valve helps to carry out purging of the HPLC system. Every time the HPLC system is switched on, it is very necessary to carry out the purging of the system. The purging of the system is carried out by rotating the purge valve in the anticlockwise direction. On opening the purge valve, the mobile phase is sucked in from the solvent reservoirs and it is flushed through the inlet tubings to remove any dissolved gases that may be entrapped in the tubings. When the HPLC system is left unused once the purging of the system is carried out the purge valve needs to be closed. The purging is carried out for a total period of 3 min with a flow rate of 8 mL/min. The typically recommended flow rate in HPLC is from 0.1 to 2 mL per minute. The flow rate in majority of the HPLC operations is kept at 1 mL/min. The minimum pressure limit is usually set to 0 or 2 bar pressure and the maximum pressure limit is usually set to 350 bar pressure whenever the pressure of the system exceeds the maximum pressure limit, the pump is automatically shut off. This helps to protect the column and the other flow components present in the HPLC system. The pump outlet is connected by a stainless-steel tubing to the rheodyne. Rheodyne sample injector is designed to deliver accurate volumes of the analyte onto the head of the column providing reproducibility of sample measurement. The Rheodyne sample injector has two positions marked on it: the load position and the inject position. Sample is injected using a hypodermic syringe through the pinhole which is present on the system with the rotatable knob in the load position. In this position, irrespective of the volume of the sample injected, 20 μL of the sample is loaded in the sampling loop which is mounted at the rear of the system, while the excess is drained out of the system when the rotatable knob is moved from the load position to the inject position. The mobile phase pumped by the high-pressure pump passes through the sampling loop and flushes the analyte under high pressure onto the head of the column.

Some precautions that we need to take into consideration during an HPLC operation to ensure smooth and reproducible measurements and also to improve the column life:

1. It is always necessary to use the required grade of chemicals in HPLC.
2. Always use solvents that are previously degassed and filtered.
3. Always use those mobile phases which are compatible with the HPLC system: for silica based stationary phases you need to ensure that you maintain the pH of the mobile phase between 2 and 8.

4. The inlet filters or the suction filters used in HPLC should be always completely immersed into the mobile phase and when the HPLC system is not in use, the suction filter should always be immersed in methanol.
5. Never allow the system to run dry.
6. HPLC columns should be always stored at room temperature.
7. Avoid any mechanical damage to the HPLC column.
8. Always spend sufficient time for column washing and column equilibration to ensure smooth HPLC operations.
9. Also, we need to ensure that we do not cause any sudden fluctuations in the mobile phase temperature, composition, and the flow rate so as to protect the column.
10. Always store the HPLC column in a methanol water system in a 70:30 ratio when not in use.

If all the abovementioned precautions are taken into consideration, we will be able to carry out very smooth HPLC operations.

The thermally stable and volatile compounds are analyzed by gas chromatography. It can be applied to certain other compounds like carbohydrates and peptides but the sample should be derivatized chemically prior to analysis so as to become volatile in nature. In HPLC, the sample (solid or liquid) is dissolved in a suitable solvent before subjecting it to the chromatographic column for separation. Certain interactions occur between the analyte or sample, column, and the mobile phase which are, namely solid–liquid adsorption, liquid–liquid partitioning, size exclusion, ion exchange, and by solute/mobile phase interactions. Smaller biomolecules which are soluble in acetonitrile/water mixtures like amino acids, sterols, steroids, carbohydrates can be separated by reversed phase chromatography. Organic solvents like acetonitrile can cause problems during the separation of proteins, as it can denature the protein or reduce its solubility. Generally, the basic instrumentation is essentially the same in each case. The outline of a typical HPLC instrument is depicted in Fig. 20.

Stationary Phases: The stationary phase in liquid–liquid chromatography is a liquid film coated on a packing material composed of 3–10 mm porous silica particles. With repeated usage of the stationary phase, the column bleeds (loses) due to the stationary phase's partial solubility in the mobile phase. By binding the stationary phase with silica particles, this bleed is prevented. The bound stationary phase's silica particles are reacted with organochlorosilane (general formula Si $(CH3)_2RCl$, where R is an alkyl or substituted alkyl group). The kind and nature of the alkyl group in the organosilane dictate the stationary phase's characteristics. As an example, the existence of a polar functional group causes the stationary phase to be polar in character. Due to the polarity of the

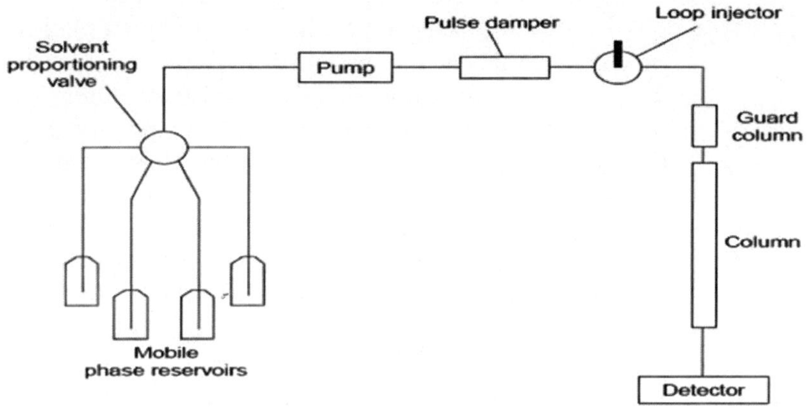

Fig. 20 Diagrammatic representation of an HPLC system

stationary phase, the mobile phase must be a nonpolar or moderately polar solvent. Normal-phase chromatography is the process of combining a nonpolar or a moderately polar mobile phase with a polar stationary phase. In comparison, reverse-phase chromatography, in which the mobile phase is more polar than the stationary phase, is the most often employed kind of HPLC. When the alkyl group in the organochlorosilane is *n*-octyl (C8) or *n*-octyldecyl (C18), the stationary phase becomes nonpolar and is suitable for reverse-phase chromatography. For reverse-phase separations, the mobile phase is often a buffered polar aqueous solution. The mobile phase's pH should be kept below 7.5 to prevent silica from being hydrolyzed when exposed to basic solutions.

Mobile Phases: The polarity of the HPLC determines the order in which the solutes get eluted. In a normal-phase separation the solute which is the least polar in nature resides proportionally less time with the somewhat a more polar stationary phase and elutes at first from the column. The nature of the mobile phase controls the retention time as a mobile phase with lesser polarity will result in longer retention time. The order of elution of the solutes in the reverse-phase separation gets reversed as the most polar solute elutes at the first. The polarity of the mobile phase is directly related to the retention time of the solute. The pH of the mobile phase is maintained between 2 and 8 by using salts like ammonium acetate, hydrogen carbonate, or phosphate at a concentration of about 20 mM. The hydrophobicity of the charged analytes can be increased by using *ion pairing* reagents at a concentration of 0.1%, as the analyte can form ion pair complex with it. Trifluoroacetic acid (TFA) an anionic pairing reagent binds with the positively charged analytes, whereas negatively charged analytes bind with tetra alkyl ammonium salts which act as cationic ion pairing reagent. Such complexes get retarded to a greater extent by the stationary phase which makes their separation easier from the unretained charged analytes alone.

Isocratic elution vis-à-vis gradient elution: If the mobile phase used is of fixed composition and contains a single component then it is referred as **isocratic elution**. A mobile phase made up of a single component is not suitable for all the solutes, it is generally observed that a mobile phase which facilitates the early elution of the solute may give rise to unacceptable longer retention time for later eluting solutes and vice versa. This problem can be addressed by altering the composition of the mobile phase as a function of time. In case of a reverse-phase separation, initially the mobile phase is relatively polar and its polarity reduces with the progress of the separation progresses. These types of separations are referred as gradient elution.

HPLC plumbing: The availability of many solvent reservoirs is an essential feature of an HPLC equipment. The mobile phase's polarity should be properly regulated to ensure adequate separation of the analyte (s). Numerous solvent reservoirs are used to modify the polarity of the mobile phase. This fluctuation is required during the gradient elution process, which involves changing the composition of the mobile phase from a weaker to a substantially stronger composition. Prior to using the mobile phase, it must be degassed to remove dissolved gases such as N_2 and O_2, as well as small particulates such as dust. When the mobile phase reaches the detector, the dissolved gases cause the formation of air bubbles, distorting the detector's signal. The most often used method of degassing is by the use of a vacuum pump or by using an inert gas (He) with a reduced solubility in the mobile phase. Filtration is used to eliminate particle matter that might block the HPLC column or tubing. The pump draws the mobile phase from the reservoirs. Generally, a reciprocating pump is utilized because it can maintain a consistent flow rate of many milliliters per minute while achieving the high output pressure necessary to drive the mobile phase down the chromatographic column.

Sample introduction: As the operating pressure employed in the HPLC is sufficiently high, so the sample is injected by the means of a **loop injector**. During loading of the sample, the loop is not in contact with mobile phase and is accessible to the atmosphere, which is then turned into the inject position once the sample is loaded. This position facilitates the mobile phase in the sampling loop which then sweeps the sample into the column.

2.9.1 Dimensions of HPLC Columns

Typically, an HPLC assembly consists of two columns: an analytical column for separation and a guard column. The guard column is put in front of the analytical column to prevent contamination. The most frequently used analytical columns for HPLC are made of stainless steel and have an internal diameter of 2.1 to 4.6 mm and a length of around 30 to 300 mm. These columns are densely packed with porous silica particles ranging in size from 3 to 10 μm and having an irregular or spherical form. Micro columns consume less solvent and provide greater signals at the detector due to the sample being diluted less. We designed open tubular micro columns with an internal diameter of 1–50 μm and a length of roughly 1 m that are free of packing material and capable of achieving column efficiencies of up to one million theoretical plates. Two issues shorten the life of an analytical column. First, irreversible solute binding to the stationary phase degrades the column's performance by reducing the amount of accessible stationary phase. Second, the sample may block the analytical column due to particle material injected with it. To mitigate these issues, an analytical column is preceded by a guard column. Guard columns are often made out of the same particle packing material and stationary phase as those of the analytical columns, but are substantially shorter and less costly. Guard columns are changed on a regular basis since they are designed to be sacrificial.

2.9.2 HPLC Detectors

Various detectors are available to monitor the HPLC separation process. The HPLC detectors, that is, UV/Vis absorption and fluorescence are based on spectroscopic measurements. The detector plots a chromatogram which is the absorbance of the analyte as a function of its elution or retention time. Instruments containing a diode array spectrophotometer generate a three-dimensional chromatogram which is the function of the absorbance at a specific wavelength and the elution time.

- The mobile phase leaves the HPLC column and fills the flow cell.
- Light from the UV or visible lamp shines through the cell.
- Light emergent from the flow cell is measured using photodiodes which produce an electric signal when exposed to light.
- The lower the intensity of emergent light, the greater the solute absorbance and larger the transmittance signal reaching the photodiodes.
- Analytes with a UV chromophore will give rise to large absorbance differences.

Quantitation is done by using Beer–Lambert law which states that absorbance is proportional to analyte concentration as follows:

A = ebc.

e = molar absorptivity (Lmol^{-1} cm^{-1}).

B = pathlength (cm).

C = concentration (mol L^{-1}).

Area of the peak is used for the quantitation. It is preferred for the tailing peaks. Peak height is useful for trace analysis or for poorly resolved peaks.

Advantages of UV-Visible detector

- They are sensitive with sensitivity in the range of 0.5–1 ng.
- Well suited in gradient elution.
- They can be used in quantitative mode.
- Compatible with a wide range of analytes.

Disadvantages

Specificity of the generated spectra is a problem (not highly specific).

Absorption dependent on the condition of the solution.

Reactive response factors need to be taken into account for quantification of the impurity.

Applications

Analysis of any UV-Visible active molecule, biological macromolecules, and small organic molecules.

Diode array detectors

They can be used for single- or multiple-wavelength detection.

The produced spectra may be retained for further analysis of the peak's purity, signal extraction, or library searching.

1. The lamps, which are composed of tungsten and deuterium, produce light with a wavelength of between 190 and 850 nm.
2. Radiation is collimated through the flow cell than a mechanically controlled slit.
3. The radiation is divided or dispersed into individual wavelengths by the holographic grating.
4. Each photodiode receives a discrete narrow wavelength band.
5. At an interval of every −12 ms, the complete spectra is recorded and stored.

Advantages

- Detection at multiwavelength.
- It can scan at a very fast speed.
- Estimation of peak purity.
- Elimination of interfering peaks.

Disadvantages

- They have lower sensitivity vis-a-vis single wavelength detector.
- Detectors are prone to lamp fluctuations.
- Peak purity is only indicative.

Applications

- Conformance of the peak purity.
- For structural elucidation or determination using the available library.
- Broad bandwidth strategies can be employed that detect any UV active components.

Fluorescence detectors provide additional selectivity as some of the analytes or solutes exhibit fluorescence or phosphorescence. These detectors have a better sensitivity as compared to UV absorbance detectors by a minimum magnitude of one order.

- The radiation is generated through the Xenon arc flash lamp.
- Radiation is reflected by the mirror onto the monochromatic excitation grating.
- The emitted radiation is dispersed and reflected by the monochromator.
- The light is split in the flow cell which then reaches the photomultiplier and reference diode.
- The noise is removed or reduced by removal of radiation below a certain wavelength by a cut-off filter prior to reaching the emission monochromator.
- The wavelength range or the wavelength of the light reaching the photomultiplier is determined by the emission monochromator.
- Photomultiplier incident photons hit the photocathode resulting in the generation of the electrons, thus amplifying or multiplying the signal.
- Optimization of the excitation and emission wavelength is of utmost importance.
- The excitation spectrum generated by the fluorometer helps in selection of the excitation wavelength.
- Type of instrument and the compound determines the optimal excitation wavelength.

Electrochemical detectors are another kind of HPLC detector that are often used. They are based on electrochemical measurements such as coulometry, voltammetry, conductivity, and amperometry. The column's effluent flows over the working electrode, which is maintained at a voltage conducive to oxidizing or reducing the analytes. The potential is maintained constant in relation to a reference electrode downstream, and the current flowing between the working and auxiliary electrodes is monitored. Electrochemical detectors detect compounds that are capable of undergoing redox reactions. Electron gain or loss is determined as the sample travels between electrodes maintained at a constant–voltage difference. Sensitivity of these detectors is in the range of 10–12 to 10–13 g/mL. Amperometry is the most often used electroanalytical detection technique. These detectors are capable of detecting functional groups such as aldehydes, ketones, phenols, amines, ethers, hydrocarbons, nitro compounds, amides, esters, olefins, halogens, quinolines, and diazo groups.

A **refractive index detector** measures the change in the refractive index of the mobile phase and is compatible with almost all compounds with the limitation being poor sensitivity. This detector cannot be used when the gradient elution is used until the components of the mobile phase have identical refractive indexes.

- Mobile phase flushing of the sample and reference cell.
- Reference cell is closed and the solvent flows only through the sample cell.
- Refractive index of the mobile phase in both cells is same.
- The elution of solute from the column into the sample cell alters the refractive index.

The light is deflected, resulting in an insufficient quantity of light reaching the diode.

Advantages

- Applicable for analytes that do not absorb radiation or wavelength in the UV/Vis range or exhibit fluorescence.
- Nonionic compounds can be detected.
- Considered as a universal detector for routine HPLC operation.

Disadvantages

- Cannot be used with gradient analysis.
- Dependent on the flow rate and temperature.
- Equilibration time required is much longer.
- Food analysis.

Evaporative light scattering detector

- The mobile phase is nebulized using a pressurized gas.
- The mobile gas is evaporated from the droplets to produce particles.
- The scattering of the light and the signal produced depends directly on the size of the particle.
- The analyte's molecular weight is not related with the output from the detector.

Advantages

- Nonvolatile analytes can be analyzed.
- Properties like UV or refractive index of the analyte or altering the eluent composition do not influence the output of the detector.
- Identifies compounds with no chromophores.

Disadvantages

- The mobile phase must be volatile.
- The detector response is a complicated function of the quantity of analyte injected.
- Analysis of smaller volatile molecules is difficult.
- Linearity of the signal with the concentration of the analyte is an issue.

Applications

- It has applications in drug discovery, especially for charged small polar species.
- It has also applications in natural product development.

A mass spectrometer is another valuable detector since it gives qualitative and structural information that may aid in identifying the analytes.

2.9.3 Choice of Buffers for HPLC Separations

Buffers to be used in HPLC should meet the following criteria:

- The pH of the buffer should be in the range of its p$K_a \pm 1$.
- At this pH, the buffer exhibits maximum buffering capacity and even a lower concentration of the buffer will give accurate results. Buffering capacity gets reduced to one-third when the pH of the mobile phase is ± 1 pH unit from the pK_a of the buffer. Incorrect selection of the buffer will demand higher concentrations of buffer solutions affecting the robustness and selectivity of the separation.
- The pH of the buffer should be adjusted first to have accuracy in pH measurement.
- Wherever possible measure buffers by weight or use a precise pipette.
- In specific, at a lower wavelength like 200 nm range, consider the UV cuts in the buffer system.
- UV cut-off of triethylamine (TEA) and trifluoroacetic acid (TFA) increases as they get degraded with time.

- Stainless steel gets corroded when exposed to citrate buffer for longer periods—ensure that columns are free from citrate buffers before storage.
- The solubility of salt buffer is decreased by modifying the counter ion in order $NH_4 < K < Na$.

Buffers Used in MS

Volatile buffers are frequently used to prevent fouling of the air pressure ionization (API) interface between the HPLC and MS instruments. It is important to keep in mind that TFA (trifluoroacetic acid) is not a buffer. It does not exhibit buffering capacity in the pH ranges which are maintained in reversed phase HPLC. It maintains the pH of the mobile phase far away from the pKa of the analytes such that the chromatographic retention or selectivity is not affected by small changes in pH. Its disadvantage is that it forms a pair with the ionized molecules of the analyte in the gaseous phase within the API interface and reduces significantly the sensitivity of MS for certain analytes. The interaction of TFA with analytes should be taken into consideration before using it. Formic acid, although also pair with the analyte ion is preferred over TFA, as the ion pair strength is low. The ion pair gets dissociated in the interphase, which allows the charged analyte in the gas phase which can be detected by MS.

2.9.4 HPLC Solvent Pumping Systems

Single Piston Pump

Most HPLC pumping devices are based around this simple design.

Components.

Eccentric cam.

Spring mounted piston.

Liquid filled chamber.

Check valves to regulate flow direction.

One disadvantage is that liquid is not delivered in a continuous manner, resulting in pulsed flow and baseline disturbances.

Dual Piston Pump

Consists of identical hydraulic chambers and piston which are operated 180° out of phase.

The major benefit is the delivery of a practically pulse free flow.

This design is often used in conjunction with a pulse dampener to provide the lowest pressure fluctuations.

Binary Pumps

- Exact mixing is required when performing gradient HPLC.
- Binary pumps have two channels or pumps each with an identical reciprocating pump connected to a low volume mixing chamber and pulse dampener.

- Solvent fraction delivered by each pump is controlled by computer software.

Quaternary Pumps

These pumps will deliver up to four solvents via a mixing device located prior to the pumps.

There is one dual piston reciprocating pump and a solenoid controlled proportioning valve located in line between the solvent degasser and pump head.

All mixing of solvents is done under low pressure.

Proportioning valve controls the volume fraction of each solvent delivered to produce the mobile phase.

Ternary Solvents

Delivers up to three solvents simultaneously.

Less commonly found in labs with isocratic, binary, or quaternary pumps being more popular.

Quaternary pups can also operate as a ternary pump.

Similar operation to the quaternary pump with the solvents being mixed under lower pressure.

2.9.5 Preparative HPLC

For various laboratory applications when the primary objective is to prepare bulk (milligrams) of the sample these columns are utilized. The column diameter of these types of columns is usually large so as to handle large sample volume injections in the HPLC. In spite of having similarity in the preparative and analytical scale HPLC separations there still occurs a number of differences.

The analytical separation aims at production of a chromatogram having well-resolved, sharp, and symmetrical peaks, along with the generation of the pure compound in sufficient quantity both economically and easily. These columns are a tube made up of stainless steel, synthetic polymers, or glass filled with microparticulate porous silica. The density of the packing material influences the separating performance of the column.

Objectives of Preparative HPLC

For any preparative HPLC it is required to maintain a proper balance between the downstream processing steps and the cost involved in the separation, time required for separation, and the operation of the available system. Preparative separations should be scaled in such a manner so that the entire process should be simple, robust, cost-effective without any loss or degradation of the product.

2.9.6 Features of Different Types of Chromatography (Table 11)

2.9.7 Comparison of HPLC, uHPLC, and FPLC (Table 12)

2.9.8 Applications of the Chromatographic Techniques in Dairy Chemistry

Isolation and Detection of Amino Acids by Paper Chromatography

Principle: Amino acids may be separated in a mixture or sample based on their solubility differences and hence their differential partition coefficients in a binary solvent system. Amino acids having a higher solubility in the mobile phase move at a relatively faster rate than those having higher solubility in the stationary phase. A solution of ninhydrin is sprayed on the dried air chromatographs to identify the separated amino acids. All amino acids develop bluish or violet color except for proline and hydroxyproline which develop yellow color.

The following is the reaction that takes place between the amino acids and ninhydrin:

Ninhydrin + amino acid → diketohydrindylidene diketohydrindamine (purple or bluish colour)

ammonium salt + carbon dioxide + ammonia + aldehyde

Materials and Reagents:

1. Hair dryer.
2. Petri plates (150 mm).
3. Developing solvent: Water:acetic acid:butanol (5:1:4).

Table 11
Features of different types of chromatography

Scale	Column I.D. (mm)	Quantity of product	Typical column length (mm)	Purpose
Analytical	4.6	1–40 mg	250	Biological materials for activity testing
Semi-preparatory	10–30	100 mg to 3 g	250	Reference compounds
Preparatory	50–70	5–10 g	250–1000	Intermediates for lab synthesis
Pilot	100–300	20 g to 5 kg	300–1000	Pharmaceutical development
Process	>300	500–1000	500–1000	Large scale production

Table 12
Comparison of HPLC, uHPLC, and FPLC

	HPLC	uHPLC	FPLC
Material	All parts of whole instrument are made of high quality stainless steel		High quality plastics or steel
Pressure	0–550 bar	Excess of 1000 bar	0–40 bar
Column	Standard analytical (4–5 mm, 10–30 cm length)	Capillary	Microbore column (1–2 mm *10–25 cm length)
Flow rate	0.010–10 mL/min		1–499 mL/min
Particle size	3–5 μm	Down to 1.7 μ	5–15 μ

4. Ninhydrin spray reagent: Dissolve 0.2 g ninhydrin in 100 mL acetone, which should be prepared fresh.
5. Standard amino acids: Prepare solutions of amino acids such as threonine, glycine, alanine, methionine, etc. (1 mg/mL in 10% isopropanol).
6. Oven set at 105 °C.
7. Sprayer.
8. Micropipette/micro syringe.
9. Whatman No. 1.

Procedure:

1. Take Whatman No. 1 filter paper with a size slightly bigger than that of the petri plate to be used in the experiment.
2. Using a lead pencil a circle of radius 1.2 cm is drawn on it. Insert the paper wick through a hole punched at the center of the paper.
3. Using a micropipette, 20 μL solution of each standard amino acid and the sample should be applied at different positions on the circle drawn on the filter paper. Ensure that the size of the spot should be small so that the developed spots do not overlay each other. Dry the filter paper.
4. The filter paper is passed over a solution of ammonium hydroxide to neutralize the amino acids and is placed over the petri plate containing the developing solvent such that the center of the paper coincides with the center of the petri dish.
5. The paper wick is inserted in the center of the paper so that its end is immersed in the developing solvent.

6. The paper should be removed when the solvent front reaches the periphery of the paper. Mark the solvent front using a lead pencil and dry the paper.
7. Spray the ninhydrin reagent on the paper and let it dry to room temperature. Transfer the paper in the oven maintained at 105 °C for 5–10 min. The colored spots, that is, purple or blue will appear on the paper. Mark the boundary of each spot with lead pencil.

Calculation:

1. Measure the distance traveled by the solvent front, sample, or the standard amino acid solution from the spot applied on the paper.
2. The R_f for the sample or the standard amino acid solution is calculated by the formula:

$$R_f = [\text{Distance moved by unknown amino acid from origin}] / [\text{Distance moved by solvent from origin}]$$

3. The calculated R_f values of the amino acids in the sample are compared with the standard R_f values of the amino acids.

Precautions:

1. The paper should not be touched with bare hands to avoid the contamination, as sweat contains amino acids in it.
2. Ensure that the spot of the sample should be as small as possible, larger spot lead to poor resolution.
3. The paper should be thoroughly dried before the spraying of the detection reagent. Wet paper may interfere with the appearance of evenly shaped compact spots.
4. The temperature of the petri dish should be properly controlled, as temperature fluctuation may result in the uneven flow of the solvent subsequently altering the R_f values.

Use of Thin Layer Chromatography to Separate and Identify Amino Acids

Principle: Separating amino acids over a thin layer of supporting media is comparable to paper chromatography (Subheading 2.1) in many respects, but offers additional benefits as discussed in section (Subheading 2.2).

Materials and Reagents:

1. Oven.
2. Sprayer.
3. Developing solvent: water:diethyl amine:acetone:n-butanol (5: 2:10:10).
4. Ninhydrin spray reagent: Dissolve 0.1 g ninhydrin in 100 mL acetone (0.1%), should be freshly prepared.

5. Standard amino acids: Make solutions of methionine, alanine, glycine, threonine, and other amino acids (1 mg/mL water).
6. Cellulose powder for TLC (300 mesh).
7. Hair dryer.
8. Micropipette/micro syringe.
9. TLC applicator.
10. TLC glass plates (20 cm × 20 cm).

Procedure:

1. To make the plates, dissolve 20 gm cellulose in 80 mL distilled water in a blender and homogenize for 1–2 min; spread on glass plates, 0.3–2 mm thick, and oven for 30 min at 100–120 °C.
2. Apply 1–5 μL of the sample or the standard on the mark drawn 25 mm from the edge of the TLC plate and allow the plate to dry.
3. Place the plates in the glass chamber containing the developing solvent.
4. Remove the plates as the solvent reaches the top edge of the plate and air-dry it.
5. Spray the ninhydrin reagent on the plate and transfer it to the oven maintained at 95 °C and observe for the appearance of purple/blue color.

Calculations and Observations

The R_f value of the sample and the standard amino acids is calculated as mentioned above and compare the calculated values with that from the reference values.

Separation of Milk Fat Components by Thin Layer Chromatography (TLC)

Principle: Thin layer chromatography can efficiently separate polar and nonpolar lipids into distinct fractions (TLC). It is a chromatographic method in which various components in a sample are separated as they pass through a very thin layer of appropriate chromatographic media, generally 0.20–0.25 mm thick, that is distributed as a uniform layer on glass plates. For this experiment, silica gel-G is employed as the chromatographic medium. Different lipids are adsorbed with varying degrees of strength into activated silica gel-G. Non-adsorbed or poorly adsorbed lipids travel with the mobile phase or the solvent, while those held firmly get retained by the stationary phase and thus travel at a relatively slower pace. This difference in their distribution between the two phases results in the separation of a sample into various components.

Materials and Reagents:

Same as for the first 7 in Subheading "Use of Thin Layer Chromatography to Separate and Identify Amino Acids".

8. Solvent A: Mixture of glacial acetic acid, diethyl ether, and hexane (1:30:70).

   Solvent B: Mixture of glacial acetic acid, diethyl ether, and petroleum ether (boiling point 250 °C) (1:20:80).
9. Spraying reagent: 50% sulfuric acid.
10. Ghee sample.
11. Standard lipids mixture: 0.2% solution of stearic acid, palmitic acid, phosphatidyl ethanolamine, lecithin, tristearin, cholesterol in hexane.

Procedure:

1. Make a slurry using 8 g silica gel-G, 22 mL distilled water, and a pastel and grinder. Pour the slurry over glass plates 20 × 20 cm in size, distribute evenly and uniformly using an applicator, and air-dry the plates. The plates are then activated by heating them to 110 °C for 20 min after they have dried.
2. Dissolve 1 mL fat in 1 mL hexane and spot about 10–20 μL on the TLC plate's marked region (2 cm from the bottom border).
3. Place the TLC plate in the chromatographic chamber containing the developing reagent, that is, A or B. The chamber should be saturated with the developing reagent, by lining the inner walls with a thick chromatographic paper or blotting paper.
4. Remove the plates from the chamber as the solvent reaches the top of the plate. Air-dry the plate and spray with 50% sulfuric acid solution and expose to 110 °C for 30 min.

Observations: The mobility of the various lipids is of the order phospholipids < monoglycerides < diglycerides < cholesterol < free fatty acids < short-chain triglycerides < long chain triglycerides.

Calculation: Calculate and compare the Rf value as discussed above.

Recovery of Proteins from Cheese Whey Using Gel Filtration

Principle: The separation occurs on the basis of the molecular weight of the compounds like salts and lactose get separated from proteins due to the lower molecular weight of the former than the latter. Salts, lactose, and other components with a lower molecular weight occurring in whey are incorporated into the gel particles, that is, Sephadex G-25 and get retained to be eluted later, while the proteins do not enter the gel particles and pass with the void volume of the column.

Materials and Reagents:

1. Deionized water.
2. Cheese whey.
3. Sephadex G-25.

4. Glass column (5 cm × 60 cm).
5. UV cord, fraction collector, peristaltic pump.

Procedure:

1. Prepare the gel column by dissolving 100 g of Sephadex G-25 (medium) in 1000 mL of distilled water for 48 h. The slurry should be packed properly to eliminate air bubbles or inconsistent packing, resulting in a 5 cm × 35 cm bed size.
2. Sample application: 50 mL of clarified whey is applied and eluted with deionized water.
3. Using a fraction collector equipped with a UV cord, collect the eluting fraction. Employ a peristaltic pump to maintain a flow rate of 150 mL/h.
4. Each drop of eluate is scanned and recorded automatically at 280 nm.
5. The whey proteins are eluted in the column's void volume portion.
6. Blue dextran is used for determining the void volume for the column while potassium chromate is used to calculate total volume.

Observations:
Void volume of the column will elute all of the whey proteins.

Using Sephadex G-25 to Concentrate Dilute Protein Solution

Principle: On addition of the dry beads of Sephadex G-25 to the dilute protein solution leads to the swelling of the beads by absorption of water. So the proteins being macromolecular in nature remain excluded from the gel beads and remain in the solution. With the absorption of water, the decrease in the volume of the solution leads to a concomitant increase in the concentration of the high molecular weight compounds like proteins without any effect on their actual quantity.

Materials and Reagents:

1. Centrifuge.
2. Quartz cuvette.
3. Spectrophotometer.
4. Sephadex G-25.
5. Deionized water.
6. α-Lactalbumin.

Procedure:

1. Dissolve 50 mg α-lactalbumin in 100 mL of deionized water. Calculate the protein concentration of the solution by monitoring the absorbance at 280 nm.

2. To the protein solution add 5 g of dry Sephadex G-25 and keep it undisturbed at room temperature for 30 min to ensure swelling of the beads.
3. The solution is then centrifuged for 10 min at 3000 rpm (503xg). Decant the supernatant in the measuring cylinder and record its volume. Take 1 mL of the solution for recording its absorbance at 280 nm.
4. Repeat **steps 2** and **3** thrice, record the volume obtained, and keep 1 mL of the sample to record the absorbance at 280 nm. Also reduce the quantity of Sephadex G-25 progressively like from 5 g in first step to 1 g in the final step.
5. Determine the absorbance (280 nm) of protein solution in the supernatant obtained after each step.
6. The increase in the absorbance of the protein solution at 280 nm after every step indicates the protein concentration in the solution.

Observations and Calculations

It is observed that after every step, though the volume of the solution decreases but the concentration of the protein (protein/mL of solution) increases.

Separation of Bovine Serum Albumin (BSA) and Blue Dextran Using Gel Filtration and Determination of Kav of Bovine Serum Albumin

Principle: The principle of separation of solutes during gel filtration is on the basis of the molecular size. The molecules with a larger molecular weight are eluted first than the molecules of lower molecular mass due to the molecular sieving effect.

Materials and Reagents:

1. 0.05 M phosphate buffer (pH 7.0).
2. Sephadex G-100 (40 μm-120 μm).
3. Potassium chromate.
4. Distilled water.
5. Blue dextran (2 mg/mL).
6. Bovine serum albumin (5 mg/mL).
7. Glass column.

Procedure:

1. Soak 18 g of Sephadex G-100 in phosphate buffer (0.05 M, pH 7.0) overnight and remove fines by decantation.
2. Pack column to give bed size of 1.6 × 80 cm and equilibrate with phosphate buffer.
3. Void and the total volume are determined by using blue dextran and potassium chromate, respectively.
4. Flow rate is maintained at 40 mL/h using the peristaltic pump.

5. A mixture of prepared BSA and blue dextran (1.5 to 2.0 mL) is applied to the column.
6. Phosphate buffer (pH 7.0, 0.05 M) is used to elute and collect the fractions.
7. The absorbance of each collected fraction is determined spectrophotometrically. For blue dextrin take reading at 540 nm and for BSA at 280 nm.

Observations and Calculations:

1. Plot the absorbance versus volume of the eluate.
2. Elution volume, V_0 of blue dextrin, total volume, V_t of potassium chromate, and V_e of BSA.
3. K_{av} is then calculated as follows:

$$K_{av} = \frac{V_e - V_0}{V_t - V_0}$$

Fractionation of Casein by Anion Exchange Chromatography

Principle: The charge on a protein depends on whether the pH of solution exceeds or is lower than the pI of the protein. Anion exchange chromatography can be used to separate acidic proteins like caseins. The disulfide bonds of the κ-casein can be reduced by 2-mercaptoethanol, while α_{s2}-casein can be alkylated with iodoacetamide and urea is used for the dissociation of the individual caseins. By carefully pooling the selected fractions, a pure preparation of β- and αs_1-casein can be produced. k-casein elutes in a wide band, indicating the presence of several glycosylated forms. Individual casein fractions are eluted in the following order: k-, β -, α_{S2}-, α_{S1}-casein using a NaCl gradient in Tris buffer (pH 8.6, containing 6 M urea).

Material and Reagents:

1. Recorder.
2. Fraction collector.
3. UV cord.
4. DEAE-cellulose (DE-52).
5. Glass column (1.6 × 75 cm) and (2.5 × 15 cm).

Stock tris-chloride buffer (0.5 M tris, 3 M NaCl, pH 8.2).

Chromatographic buffer: (pH 8.6, 0.005 M tris, 0.03 M NaCl, 6 M urea): In around 3 l of water, dissolve 1.8 kg of urea, then add 50 mL tris-chloride buffer. and make up to 5 L with water. Filter it with Whatman Filter paper no. 1 and then pass through a short DEAE-cellulose (DE-52) glass column (2.5 × 15 cm) to remove isocyanate ions. Store the buffer in a dark brown container.

The ion exchanger (DEAE-cellulose) is prepared as follows:

1. For 30 min, suspend DEAE-cellulose (DE-52) in a 15-volume 0.5 M HCl solution.
2. Then, using distilled water, wash the gel over the Buchner funnel until the pH reaches 4. After 30 min, the resin is suspended in 15 volumes of 0.5 M NaOH.
3. The resin is then washed with distilled water to a pH of 8.0 and left to stand.
4. The heavier particles sink to the bottom, while the lighter particles float to the surface and are removed by decantation.
5. Finally, the slurry is equilibrated against chromatographic buffer and vacuum degassed prior to packing the column.

Preparation of samples:

1. Dissolve 500 mg acid casein in 40 mL imidazole-HCl–urea buffer (pH 7.0, 0.01 M imidazole, 4 M urea); add 0.04 mL 2-mercaptoethanol and stir at 40 °C for 1 h.
2. Add 375 mg iodoacetamide and mix for 30 min.
3. Finally, add 0.1 mL 2-mercaptoethanol to neutralize any remaining iodoacetamide.
4. The alkylated casein solution is dialyzed many times against a tris-chloride-urea buffer (pH 8.6, 0.005 M tris, 0.03 M NaCl, 6 M urea) to establish full equilibrium with the buffer.

Procedure:

1. Fill the column with DEAE-cellulose slurry at a flow rate of 60 mL/h, generating a bed of 1.6 × 60 cm equal to a 120 mL bed volume.
2. Stabilize the exchanger by introducing two column volumes (CV) of chromatographic buffer at a flow rate of 40 mL/h.
3. Load the column with 20 mL of alkylated casein solution containing approximately 250 mg proteins.
4. Wash with 1CV chromatographic buffer.
5. Elute using a 3CV gradient with a salt concentration of 60–250 mM in chromatographic buffer.
6. Collect the eluate at a flow rate of 40 mL/h in 5 mL fractions while constantly monitoring and recording absorbance at 280 nm.
7. Dialyze the pooled peak fractions against distilled water numerous times and lyophilize.

Principle: Lactoferrin can be readily extracted from colostral whey using immobilized metal affinity chromatography. This is because lactoferrin has a high affinity for chelated metal ions. Following epoxy activation, a metal chelator such as iminoacetic acid is linked

Lactoferrin Isolation from Colostral Whey Using Affinity Chromatography with Immobilized Metal Chelates

to Sepharose 6B. Biscarboxymethyl amino groups are connected to Sepharose 6B through a long stable hydrophilic spacer arm. In the column, the gel is activated by washing with a solution containing a suitable metal ion, such as Cu^{2+}. Due to the presence of exposed histidine, cysteine, and tryptophan residues, the protein binds to a metal chelate—Sepharose. Binding is dependent on pH and compounds are often desorbed by lowering the eluent's pH.

Material and Reagents:

1. UV cord.
2. Copper sulfate.
3. Chelating Sepharose 6B.
4. Buffer A: Tris acetate buffer (0.05 M, pH 8.2 containing 0.5 M NaCl).
5. Buffer B: Tris acetate buffer (0.05 M, pH 2.8 containing 0.5 M NaCl).
6. Peristaltic pump.
7. Fraction collector.
8. Glass column (1 cm × 20 cm).

Preparation of sample:

1. Warm 100 mL of colostrum to 400 °C.
2. Centrifuge colostrum at 5000 × *g* for 10 min at 400 °C to separate the fat content.
3. Maintain the centrifuge tubes at 40 °C for 1 h.
4. Using a spatula, scrape away the top fat layer and collect the skimmed colostrum.
5. Bring the skim colostral sample to room temperature by diluting it 1:2 with water.
6. Slowly add 1 M HCl while stirring continuously until a pH of 4.6 is obtained.
7. Allow the clear precipitates of casein to rest at room temperature for 30 min to ensure complete separation.
8. Using filter paper, separate the precipitated casein curd from the whey.
9. Separate the clear whey portion from the rest.
10. Equilibrate the whey sample overnight by dialysis against buffer A.

Procedure:

1. Pack the column to a bed height of 16 cm with chelating Sepharose 6B gel that has been equilibrated with buffer A.

2. Add Cu^{2+} to the gel by running 1 column volume of copper sulfate solution (1 mg/mL) across it.
3. Load the column with 70 mL of the whey sample.
4. Rinse the column with buffer A until the eluate absorbance is less than 0.02 at 280 nm.
5. Using buffer B, elute the lactoferrin and collect the separated protein.
6. Using SDS-PAGE, determine the purity of lactoferrin (Experiment no. 12).

Gas-Liquid Chromatography Analysis of the Fatty Acid Content of Milk Fat

Principle

The study of the fatty acid content of fats and oils using gas-liquid chromatography (GLC) has become common. Milk fat is an exception due to its greater complexity than the majority of other natural fats. The study of short-chain fatty acids, beginning with butyric acid, presents two difficulties. The first issue arises from the methyl esters of butyric and caproic acids' high volatility and water solubility. It is physically impossible to avoid the loss of significant amounts of these fatty acid methyl esters. The second issue with fatty acids in milk fat is that the short-chained esters are crowded in the first section of the chromatogram, making quantitative measurement of the peak regions difficult. To prevent the loss of methyl butyrate, methyl esters of fatty acids are immediately introduced into the GLC column without any water washing or solvent evaporation. The temperature programming results in a greater spacing between all peaks, which simplifies and improves quantitative assessment of peak regions.

Materials and Reagents

1. Milk fat.
2. Freeze-drying tubes.
3. Fatty acid methyl esters standard.
4. Absolute methanol.
5. Sodium metal.
6. GLC assembly.

How to proceed:

Preparation of methyl ester

To be analyzed, 0.2 g of milk fat is sealed in a freeze-drying tube with 0.4 mL of 0.025 N sodium methylate produced by dissolving sodium metal in anhydrous methanol. The sealed tube is put in an oven set at 80 °C for 1 h, as indicated by the shift from a two-phase to a one-phase system.

Conditions for GLC Operation:

1. Fat is analyzed for fatty acid composition using a double column GLC (Amil Nucon, 5700, Japan) equipped with a flame ionization detector, temperature programming module, and recorder (Philips), as well as a 72 0.25-in. DEGS glass column.
2. GLC should be used in accordance with the Operator's Manual under the following conditions:
   (a) FID module control: 8× attenuation and 1000 sensitivity.
   (b) Module for temperature regulation.
   (c) Detector temperature: 220 °C.
   (d) Carrier gas: Nitrogen at a flow rate of 30 mL/min.
   (e) Fuel: Hydrogen gas at a rate of 35 mL/min.
   (f) Air: Compressed air at a flow rate of 330 mL/min.
   (g) Recorder current: 10 mV.
   (h) Chart speed: 10 mL per minute.

Sample Application:

1. Using a micro syringe, inject 0.4 μL of the esterified sample into the preconditioned GLC column at 70 °C. When the first peak of butyric acid appears, increase the temperature to 150 °C until the lauric acid peak appears (C12). Then increase the temperature to 180 °C and hold the column at 200 °C until the stearic acid (C18) and oleic acid (C18:1) peaks appear.
2. After all of the sample components have exited the column, inject the standard methyl esters individually and record the retention duration of esters of various fatty acids.

Calculations:

1. Identify the individual peaks in the samples by comparing their relative position or retention time to those of the standards esters, which are ordered by carbon atoms and degree of unsaturation for the same number of carbon atoms in fatty acids.
2. Calculate and compare the peak area of each acid using the formula base/2× peak height. The amount of fatty acid can be determined by the amount of sample or standard injection and attenuation.

Determination of Fatty Acids of Ghee Using GC-MS/MS (ISO 15884:2002/IDF 182:2002, ISO 15885:2002/IDF 184:2002)

Milk fat is composed of around 400 distinct fatty acids, making it the most complex natural fat available. The fatty acid content of milk fat is measured using gas-liquid chromatography as methyl esters of fatty acids. Saponification is a classical process for producing fatty acid methyl esters. It includes the conversion of fat or oil to methyl esters and alcohol in the presence of an aqueous alkali. Saponification of milk fat is accomplished by adding a solution of potassium hydroxide (KOH) or sodium hydroxide (NaOH) to the milk fat. After a certain amount of time has passed, the mixture is neutralized with crystalline sodium hydrogen sulfate to prevent saponification of preformed esters. The fatty acid composition of milk fat is given as a mass fraction in percent (as grams of individual free acids per 100 g of total fatty acids (free acids)).

Glassware and Apparatus

Balance, centrifuge tube (15 mL capacity), micropipettes (capacity of 10 μL and 1000 μL), vortex mixer, centrifuge (capable of operating at 350 $g \pm 50$ g). GC vials (2 mL capacity).

Chemicals

1. Transesterification reagent: 2 M methanolic solution of KOH or sodium methoxide ($NaOCH_3$). Dissolve 11.2 g of KOH in 100 mL of methanol and mix well. Alternatively, dissolve 10.8 g of $NaOCH_3$ in 100 mL of methanol and mix well.
2. Sodium hydrogen sulfate monohydrate.
3. Hexane.
4. Methanol.

Procedure:

1. Preparation of methyl esters.

   Weigh, to the nearest 5 mg, 100 mg of the prepared test sample in a 15 mL centrifuge tube. Dissolve the test portion of sample in 5 mL hexane and mix well. Add 0.2 mL of the transesterification reagent and cap the tube. Mix the contents of the tube vigorously with the vortex mixer for 1 min. After an additional reaction time of 5 min, add 0.5 g of solid sodium hydrogen sulfate and mix again. Place the centrifuge tubes with the test portion in the centrifuge and centrifuge for 3 min at 350 $g \pm 50$ g at room temperature. After centrifugation, take 1.5 mL aliquot from the obtained clear supernatant of the test portion in the GC vials for the gas-liquid chromatographic analysis.

2. GC Determination.

   Analysis of fatty acid is carried out by using Shimadzu GC-MS/MS fitted with TQ-8030 triple quadrupole detector. Switch on the instrument and open the GC real time analysis software. Put the GC vials containing sample into autosampler (AOC 20i) rack. In the real time analysis software, create batch table for running the sample by inserting vial number, method file, and tuning file. Q3 scan mode which provides qualitative information of sample molecule, is selected for the analysis of fatty acid methyl ester. Then after, one microliter of sample dissolved in *n*-hexane is injected in column with autosampler (AOC 20i) through automatic syringe. The GC oven program starts with initial temperature of 100 °C (hold with 5 min), then rise temperature at the rate of 4 °C/min to 240 °C (hold with 20 min). After a solvent delay time of 8 min, mass spectra are recorded over the whole range and a total ion chromatogram is created. Peak integration is carried out in the total ion chromatogram and spectrum of determined peaks is registered into spectrum process table. The fragmented spectra of analyte molecules are matched with reference spectra pre-registered in NIST library. Quantification of FAME sample is done by area normalization method in which amount of sample components are estimated based on area/height or area of the component's peak relative to the total area multiply by 100 (% composition). Following GC-MS conditions are maintained for the analysis.

3. GC conditions:

a) Column	:	Rtx 2330 (CP Sil-88; 100 m × 0.25 mm i.d.
		a. × 0.20 μm, Restek, Bellefonte, PA, USA)
b) Oven temperature	:	100°C
c) Injector temperature	:	230°C
d) Column flow	:	1.3ml/min
e) Split ratio	:	80:1

4. MS conditions:
   (a) Ion source temperature: 220 °C.
   (b) Interface temperature: 240 °C.

   Gas Chromatography has been used to detect soyabean oil and buffalo body fat in ghee based on the differences in their fatty acid composition [5]

References

1. Wang Y, McCaffrey J, Norwood DL (2008) Recent advances in headspace gas chromatography. J Liq Chromatogr Relat Technol 31(11–12):1823–1851
2. https://www.chromacademy.com/
3. Kumar A, Upadhyay N, Gandhi K, Kumar A, Lal D, Sharma V (2013) Reverse-phase thin layer chromatography of Unsaponifiable matter of ghee for detecting adulteration with soybean oil and buffalo depot fat. Indian J Dairy Sci 66(6):496–501
4. Kumar A, Upadhyay N, Gandhi K, Lal D, Sharma V (2013) Detection of soybean oil and buffalo depot fat in ghee using normal-phase thin layer chromatography. Indian J Dairy Sci 66:4
5. Kumar A, Upadhyay N, Padghan P, Kumar A, Lal D, Sharma V (2015) Detection of vegetable oil and animal depot fat adulteration in anhydrous milk fat (ghee) using fatty acid composition. MOJ Food Process Technol 1(3): 00013
6. Mann B, Sangwan R, Kumar R (2008) Research techniques in Dairy Chemistry. NDRI, Karnal
7. Nielsen SS (2017) Introduction to food analysis. Food analysis. Springer, New York, pp 3–16
8. Pearson D (1976) The chemical analysis of foods. Longman Group Ltd, London
9. Wilson K, Walker J (2010) Principles and techniques of biochemistry and molecular biology. Cambridge University Press, Cambridge
10. https://www.wikiwand.com/en/Thermal_conductivity_detector
11. https://www.chromatographyonline.com/view/electron-capture-detectors
12. Manz A, Pamme N, Iossifidis D (2004) Bioanalytical chemistry. World Scientific Publishing Company
13. Goldsmith JG (2000) Modern Analytical Chemistry, 1st Edition (Harvey, David). J Chem Educ 77:705. https://doi.org/10.1021/ed077p705.2
14. https://www.slideshare.net/RanjithR70/chromatography-ranjith

Check for updates

Chapter 3

Centrifugation

Abstract

The technique of separating substances under the influence of centrifugal force is termed as centrifugation. Particles get separated from the solution on the basis of their density, shape, size, medium viscosity, and the speed of the rotor of the centrifuge. A centrifuge is a machine which rotates around a fixed axis, in which the force acts perpendicularly to the axis of the spin, i.e., in an outward direction. Gravity and the centrifugal force generated in the centrifuge form the basis for the separation of different fractions. This chapter covers the principle of centrifugation, types of centrifuge, their instrumentation, types of centrifugation process, and practical application of centrifugation process applied to milk. The design and working of various types of centrifuges like desktop clinical centrifuges, high-speed centrifuges, microcentrifuge, vacuum centrifuge/concentrators, and ultracentrifuges have been discussed. Three types, namely ultracentrifuge, high-speed and low-speed centrifuge, have been compared. Types of rotors used in centrifuges like fixed angle rotors, vertical tube rotors, and swinging-bucket rotors have also been discussed. Preparative centrifugation methods like differential gradient centrifugation, density gradient centrifugation which included rate-zonal centrifugation and isopycnic centrifugation, and analytical centrifugation methods have been explained.

Keywords Centrifugation, Desktop clinical centrifuges, High-speed centrifuges, Microcentrifuge, Vacuum centrifuge/concentrators and ultracentrifuges, Differential gradient centrifugation, Density gradient centrifugation, Rate-zonal centrifugation, Isopycnic centrifugation, Analytical centrifugation

The technique of separating substances under the influence of centrifugal force is termed as centrifugation. Particles get separated from the solution on the basis of their density, shape, size, viscosity of the medium, and the speed of the rotor of centrifuge. A centrifuge is a machine which rotates around a fixed axis, in which the force acts perpendicularly to the axis of the spin, that is, in an outward direction. Theodor Svedberg developed the first analytical ultracentrifuge in 1920. Gravity and the centrifugal force generated in the centrifuge form the basis for separation of different fractions.

1 Principle

When an object moves in a circular motion having a constant or steady angular velocity, it experiences an outward force, *F*. This

Kamal Gandhi et al., *Advanced Analytical Techniques in Dairy Chemistry*, Springer Protocols Handbooks,
https://doi.org/10.1007/978-1-0716-1940-7_3,

force depends on the radius of rotation, r (cm) and the angular velocity in radians, ω.

$$F = \omega^2 r \tag{1}$$

F is expressed as the earth's gravitational force and is referred as the relative centrifugal force, RCF/number times g ($g = 980$ cm/ s^{-2}).

$$\mathrm{RCF} = \omega^2 r/980 \tag{2}$$

"Revolutions per minute (RPM)" is used to express these relationships. The speed at which a centrifuge operates is given in terms of RPM. RPM can be converted to radians using the formula.

$$\omega = \pi(\mathrm{RPM})/30 \tag{3}$$

Then

$$\mathrm{RCF} = 11.17r(\mathrm{RPM}/1000)^2, \quad r = \text{radius in cm} \tag{4}$$

$$\mathrm{RCF} = 28.38R(\mathrm{RPM}/1000)^2, \quad R = \text{radius in inches} \tag{5}$$

The position of a certain particle in the sample container determines its effective radial dimension which varies between r_{min} and r_{max}. The centrifugal force which acts at the bottom and at the top of the container differs by nearly two folds. The force experienced by a particle depends on its position from the axis of rotation, that is, force acted on the particle is directly proportional to its distance from the axis of rotation. The relative centrifugal field is calculated from the average radius of rotation (r_{av}) of a column in a tube. In case of a partially filled container of a swinging bucket and with a fixed angle rotor, the r_{min} is effectively increased, thus resulting in quicker sedimentation, as they need to travel a reduced path length. For a spherical particle, the separation or sedimentation rate depends on its density, applied centrifugal force, particle radius, and viscosity of the medium. This relationship is also called as Stokes' law [1–3].

$$U = \frac{2}{9} r_{\mathrm{p}}{}^2 \frac{\rho_{\mathrm{p}} - \rho_{\mathrm{m}}}{\eta} \times r\omega^2 \tag{6}$$

where

U = sedimentation rate or velocity of sphere.

ρ_{p} = density of the particle.

ρ_{m} = density of the suspending medium.

r_{p} = radius of the particle.

η = viscosity of the suspending medium.

$r\omega^2$ = centrifugal acceleration.

$2/9$ = shape factor constant for a sphere.

If the density of medium is equal to the density of the particle, then the sedimentation rate will be zero. Apart from these two factors, size also greatly affects the sedimentation rates. Thus, it is clear from the above equation that the particles having similar densities can also differ in their sedimentation rates or velocity by virtue of their differences in their size.

Two forces, that is, buoyant and frictional force which act on a particle suspended in the medium, counteract the centrifugal force acting on it.

1.1 Buoyant Force

The upward force acting on the particles which is equal to the weight of the liquid displaced by them.

1.2 Frictional Force

Force acted on the particles by virtue of their motion through the solution.

This viscous force acting on a sphere of radius "r" moving with a velocity "v" in a liquid of viscosity "η" is given by Stokes' law.

$$F = 6\pi\eta rv \tag{7}$$

Stokes' law is valid only when the liquid possesses laminar flow. The frictional drag acting on a sphere is proportional to its radius, viscosity of the liquid, and the velocity with which the liquid flows.

1.3 Derivation of Stokes' Law

Consider a fat globule having radius "r" and density "ρ_p" moves with a velocity "v" in a liquid having density "ρ_m" and viscosity "η." The forces acting on the fat globule are gravitational force $F_G = 4/3\pi r^3\rho_p g$, frictional force or viscous drag force due to flow $F_D = 6\pi\eta rv$, and the force due to buoyancy $F_B = 4/3\pi r^3\rho_m g$. The rise of the fat globule continues until the net force acting on it equals to zero, that is, the resultant of all the three forces becomes zero. Henceforth, the sphere continues to move at a constant rate termed as the terminal velocity. Putting the values of $F_D = 6\pi\eta rv$ and $F_G = 4/3\pi r^3\rho_p g$ in the equation $F_D + F_B = F_G$, We will obtain the equation 8 as given below.

$$V_t = 2r^2\left(\rho_p - \rho_m\right)g/9\eta \tag{8}$$

Replacing "g" with centrifugal force "$r\omega^2$," we will obtain:

$$U = \frac{2}{9}r_p{}^2\frac{\rho_p - \rho_m}{\eta} \times r\omega^2 \tag{9}$$

Sedimentation or separation of the particles in a centrifuge occurs when the centrifugal force overcomes the frictional and buoyant force, resulting in the movement of the particles away from their axis of rotation. In a laboratory centrifuge, the low density or lighter substances rise to the top of the sample tube, while the heavier particles sediment at bottom of the tube, this separation of the particles occurs by the virtue of radial acceleration acting on the entire particles.

Factors which have an influence on the centrifugation include the following:

- Speed of rotation.
- Distance of particles displacement.
- Viscosity or temperature.
- Density of the solute and the solvent.

1.4 Sedimentation Coefficients

Centrifugation works on the principle of sedimentation. The sedimentation rate or the rate of sedimentation is generally termed as the rate at which the particle having a certain shape and size travels to the bottom of the centrifuge tube due to centrifugal force. The rate of sedimentation depends on the factors like viscosity of the solution, shape, and size of the particle(s). Mathematically, it can be represented as the ratio of the sedimentation velocity (v_t) of a particle to the gravitational (centrifugal) force causing it to sediment. Theodor Svedberg created this mathematical expression. The values of "S" ranges between 10^{-13} and 10^{-11} s, where 1 Svedberg (S) = 10^{-13} s [4].

2 Classification of Centrifuges

2.1 Desktop Clinical Centrifuges

Desktop clinical centrifuge is the most simple and cheapest form of centrifuge available till date (Fig. 1). Substances which can sediment rapidly (yeast cells, blood cells, bulky precipitates from certain reactions) or to fractionate smaller quantities of such substances, these centrifuges are used. They can operate up to speed not more than 3000 rpm at ambient temperature.

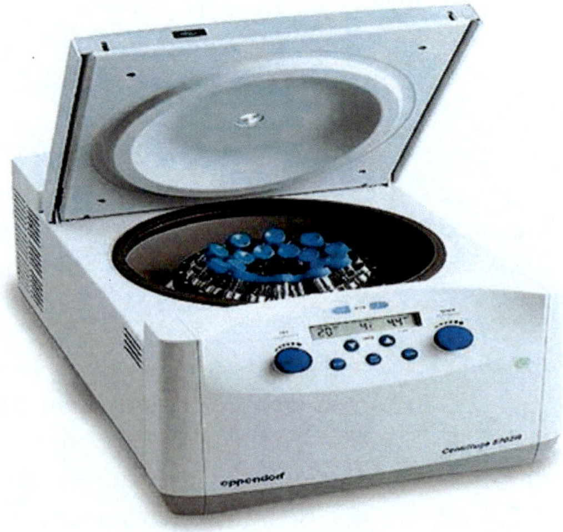

Fig. 1 Desktop centrifuge [5]

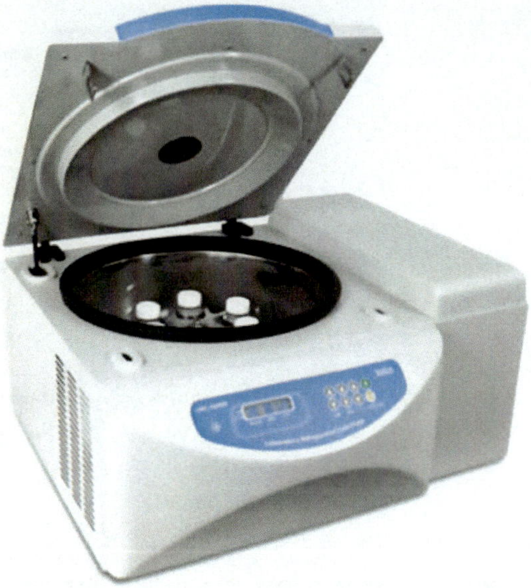

Fig. 2 High-speed centrifuge [6]

2.2 High-Speed Centrifuges

These types of centrifuges (Fig. 2) can be operated up to a speed of 20×10^3 to 25×10^3 rpm. Experiments aiming at bulk preparative applications are performed with them. These instruments are equipped with refrigeration system which ensures that the heating of the rotor chamber can be prevented. They are classified into two categories. One, a simple high capacity and a continuous flow type of centrifuge. A steam turbine drive unit or a very fast electric motor is connected to a long tubular rotor through a *V* belt assembly or a cork clutch. This type of centrifuge is mainly used for harvesting bacterial or yeast cells from large cultures having volumes ranging from 5 to 500 L. Another variant of this centrifuge is a high-speed centrifuge which is equipped with a lower capacity of the refrigeration unit. The temperature of the rotor chamber is maintained between 0 and 4 °C, through a coarsely controlled thermocouple located at the bottom of the chamber. Compared to the clinical instrument, speed is controlled in a much more superior manner in these instruments. Also these types of centrifuges are equipped with a brake known as a breaking device, which helps to decelerate the rotor at the end of the centrifugation process. Such instruments are used to separate cells, cellular debris, collection of microorganisms, immune precipitates, ammonium sulfate precipitates, large cellular organelles, and for preparative applications. Subcellular organelles like lysosomes, mitochondria, nuclei, etc. can also be isolated. They cannot effectively sediment viruses and smaller organelles (e.g. ribosomes).

Fig. 3 Microcentrifuge

2.3 Microcentrifuge

Separation of smaller quantities to the tune of 0.5 to 2 mL is accomplished by using microcentrifuge (microfuge) (Fig. 3). These centrifuges are operated at 12×10^3 to 13×10^3 rpm for separating cell organelles like deoxyribonucleic acid, nuclei, and phenol extraction. Sample tubes smaller in size than the usual centrifuge tubes are used in microfuge. Microfuges are also equipped with adapters which make it adaptable to be used with larger tubes along with the smaller ones. For processing of the temperature sensitive samples, temperature controller is also available with microfuges.

2.4 Vacuum Centrifuge/ Concentrators

Vacuum centrifuge (Fig. 4) is used for concentrating the sample by evaporating the liquid or solvent from the biological or chemical sample(s) by utilizing a combination of vacuum, centrifugal force, and heat. These centrifuges have a high sample throughput, as around 148 samples can be processed in it at a time. These are used when the samples generally contain large number of solvents. The solvent evaporation and prevention of the solvent bumping is accomplished in the rotary evaporator. The basic principle behind the working of this type of centrifuge is that lowering of the pressure in the chamber leads to the reduction in the boiling point of the samples, thus resulting in the evaporation of the solvents leading to the sample concentration.

2.5 Ultracentrifuge

These types of centrifuges have now widened the area of research in most of the fields, as these instruments are capable of attaining speeds up to $50 \times 10^4\ g$ (75×10^3 rpm, $r = 8$ cm). It facilitates the fractionation of subcellular organelles and macromolecules in their native form which could only be observed in electron micrographs previously. Successful separation of smaller molecules like viruses, proteins, and ribosomes can be achieved. The heat

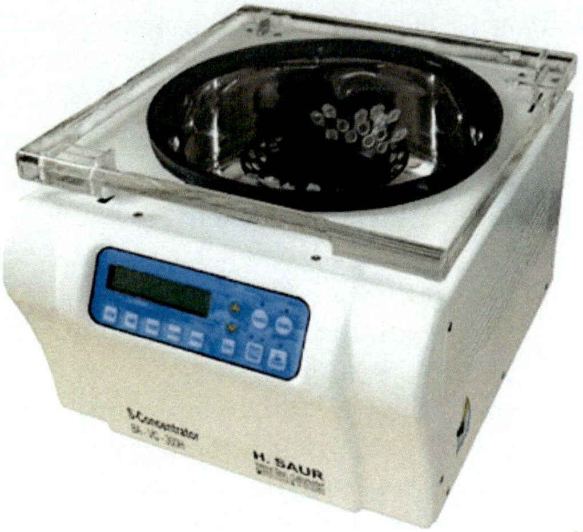

Fig. 4 Vacuum centrifuge/concentrators [7]

Fig. 5 Ultracentrifuge [8]

produced during the spinning of the rotor of the centrifuge is maintained by the refrigeration system in these instruments. Experiments related to both preparative and analytical works can be performed with ultracentrifuge. Preparative centrifugation means that process of centrifugation carried for the isolation and purification of the components while analytical centrifugation is done for the characterization. (Fig. 5)

2.6 Instrumentation of an ultracentrifuge:

2.6.1 Drive and Speed Control

The drive assembly in most instruments consists of a motor which is linked by means of a precision gear box to the rotor spindle. The small diameter of the shaft allows it to flex when spinning, thus accommodating a small amount of rotor imbalance without vibration or spindle damage. A rheostat can be used to select the speed at which a rotor spins and can be operated which can be monitored using a tachometer. In order to ensure that the rotor is operated below its maximum rated speed, a second overspeed system is fitted with the speed control system.

2.6.2 Temperature Control

Ultracentrifuge has a more sophisticated temperature monitoring system than other types of centrifuges. An infrared radiometric sensor in the ultracentrifuge which is located under the rotor monitors the temperature of the rotor directly, thus ensuring a more responsive and accurate temperature control, while in other high-speed centrifuge only the rotor chamber is monitored by the temperature control device.

2.6.3 Vacuum System

Incorporation of the vacuum system in the ultracentrifuge differentiates it significantly from other high-speed centrifuge. The heat generated due to the friction between the spinning rotor and the air is significantly lower at speeds below 15,000 to 20,000 rpm vis-à--vis at speeds above 40,000 rpm. This heating is eliminated by sealing the rotor chamber and is evicted using two types of pumping systems: a diffusion pump and a mechanical vacuum pump.

2.6.4 Rotors

Apart from the availability of a large variety of rotors, they are classified as angle and swinging bucket. The material used for their construction depends on their speed of operation, with those to be used for lower speeds are composed of aluminum alloys, while those for higher speeds are made of titanium. The angle rotors made up of a solid piece of metal consist of 6 to 12 holes machined at an angle between 20° and 45°. Such rotors are mostly used in applications involving pelleting or complete sedimentation of a constituent. The most important feature of such rotors is that they have a large capacity. The other type of rotors, that is, swinging-bucket rotors, has a rotor to which three to six freely moving buckets hang. At rest, these buckets remain hanging in a vertical position while under the influence of centrifugal force, they swing at right angle to the horizontal at a speed of 200 to 800 rpm. Table 1 presents the differences between the low speed, high speed, and ultracentrifuges.

Table 1
Types of centrifuges and applications

Characteristics	Type of centrifuge		
	Ultracentrifuge	**High speed**	**Low speed**
Range of speed (rpm)			6000
Maximum RCF (g)	6,00,000	50,000	6000
Refrigeration	Yes	Yes	Optional
Pelleting of cells	Yes	Yes	Yes
Pelleting of nuclei	Yes	Yes	Yes
Pelleting of organelles	Yes	Yes	No
Pelleting of ribosomes	Yes	No	No
Pelleting of macromolecules	Yes	No	No

3 Types of Rotors

Centrifuge rotors are categorized into three types: vertical rotors, swinging-bucket rotors, and fixed angle rotors. All three are designed keeping in view of the following factors:

1. Type of centrifugation.
2. Speed.
3. Volume range.

For low speed centrifugation, rotors experience lower degree of stress, thus perspex, steel and brass can be used for their construction, while alloy of titanium or aluminum is used for making rotors to be used in high speed centrifugation, as they experience a higher degree of stress. In order to reduce the corrosive effects of chemical reactions or due to solute or solvent, the rotors are covered with a protective layering. Swinging rotors and fixed-angle type of rotors are mainly used in low speed, high-speed floor model, and bench top centrifuge, while in ultracentrifuge vertical rotors are used.

3.1 Fixed Angle Rotors

Tubes are positioned at an angle of 14 to 40° to the vertical (Fig. 6). The particles move rapidly outwards and travel a short distance. These rotors are used for differential centrifugation. With the acceleration and deceleration of the rotor automatically reorients the tube.

3.2 Vertical Tube Rotors

These rotors are held vertically parallel to rotor axis (Fig. 7). The time required for the separation of the particles is lesser as they tend to travel shorter distance. One major disadvantage is that the pellet tends to fall back into solution again at the end of centrifugation.

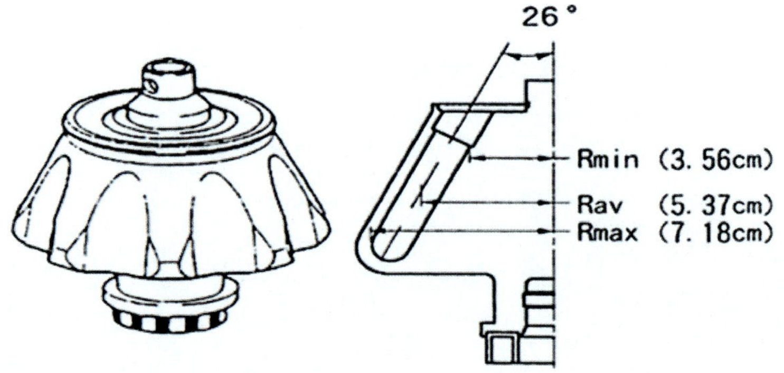

Fig. 6 Fixed angle rotor [9]

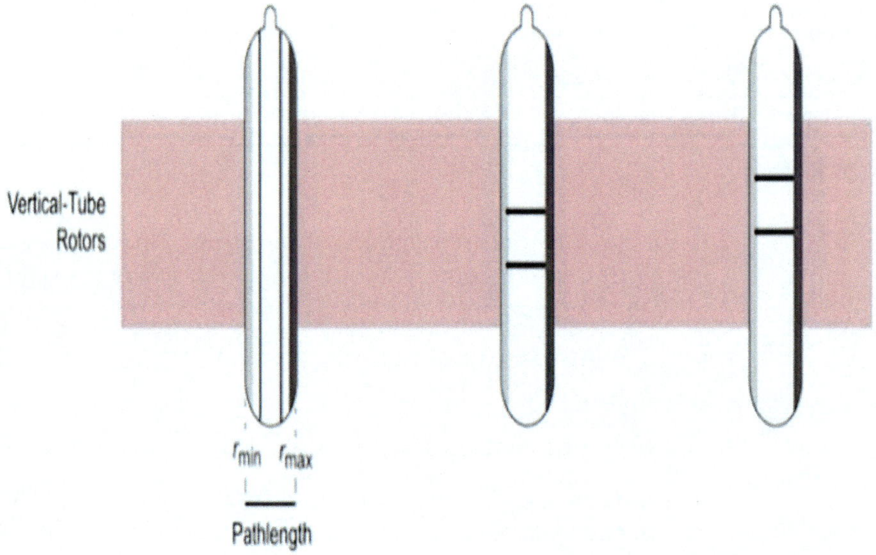

Fig. 7 Vertical angle rotor [8]

3.3 Swinging-Bucket Rotors

As the rotor gets accelerated, this type of rotor orients into the horizontal position (Fig. 8), as a result the sample has to travel a longer distance, resulting in the better separation like in density gradient centrifugation. These rotors are generally used for density-gradient centrifugation. Removal of the supernatant is much easier as it does not disturb the pellet.

4 Preparative Centrifugation

It is employed for the isolation of the biological material for further biochemical investigations. Depending on the separation of the suspending medium, this type of centrifugation is divided into two categories, namely density gradient centrifugation for density

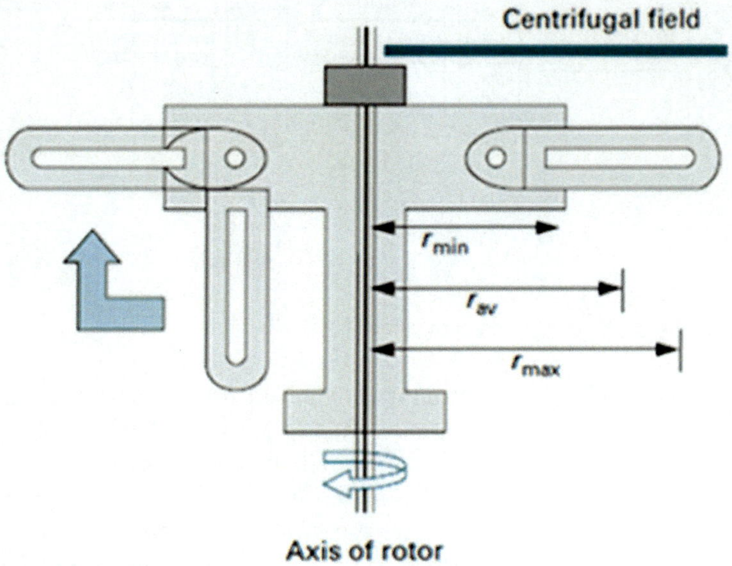

Fig. 8 Swinging-bucket rotors [8]

gradient medium and differential centrifugation for homogenous medium.

4.1 Differential Gradient Centrifugation

It is the widely used type of the centrifugation. Biological sample like liver is firstly homogenized at 32 °C in a sucrose solution containing buffer. The obtained homogenate is then transferred into the centrifuge tube and is spun at a constant speed. The homogenate then gets separated into two layers, that is, a supernatant and a pellet at the bottom. The supernatant is transferred into another centrifuge tube and subjected to higher speed in the progressing steps. The size of the particles forms the basis of the separation, as smaller particles sediment at a slower rate than their counterparts. Subjecting the supernatant to a series of increasing speed or *g*-force, more purified samples are obtained. This technique is generally used to obtain a partially pure separation of the macromolecules, cell organelles, and for simple pelleting (Fig. 9).

Despite of poor yield, differential centrifugation is still widely used for isolating intracellular organelle from tissue homogenates, which is generally attributed to the following factors:

- Time economy.
- Convenience.
- Relative ease.

4.2 Density Gradient Centrifugation

Purification of membranes, cells, and cellular organelles like ribosomes, etc., is done by density gradient centrifugation. By overlaying lower concentrations in the centrifugal tube, the density

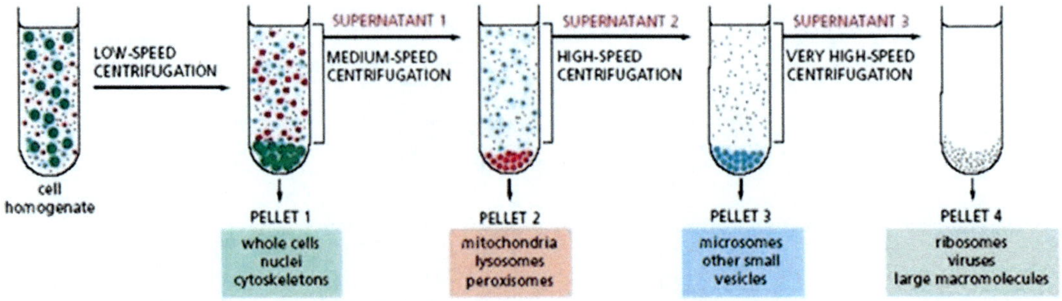

Fig. 9 Differential gradient centrifugation [10]

gradient is achieved using sucrose. Particles of concern are positioned on the top of the gradient in ultracentrifuges. Particles move across the gradient until they reach a point where their density equals the density of the sucrose surrounding them.

The fraction is then removed and analyzed. This method is suitable for purification of macromolecules and subcellular organelles. Density gradient is created by placing the heaviest layer of the gradient media at the bottom followed by overlaying lighter layers on it with the lightest being on the top, generally sucrose is used to create the gradient. Gradient materials used are glycerol, silica sols, sucrose (66%, 5 °C), polyvinylpyrrolidone, sorbitol, ficol (high molecular weight sucrose polymer and epichlorohydrin), Cs acetate, and cesium chloride (CsCl).

4.2.1 Rate-Zonal Centrifugation

This technique is also termed as gradient or band centrifugation. Separation by this method is based on the principle of sedimentation coefficient (i.e. movement of the sediment through the liquid medium). The gradient is created with sucrose in a test tube with the gradient increasing at the bottom of the test tube. The sample is placed on the top of the gradient and subjected to centrifugation (Fig. 10). The sample gets separated into various zones in the gradient, that is, the faster sedimenting particles move ahead of the slower ones. The separation of the particles depends on the sedimentation coefficient and a hole is created at the bottom of the tube in order to separate the sedimented fractions. Prerequisites for a proper rate-zonal centrifugation:

- The sample solution should have a density lower as compared with the lowest density portion of the gradient.
- Too long runs should be avoided, as the particles tend to pellet at the base of the tube.
- The sample particle should have a density greater vis-à-vis the highest density portion of the gradient.
- The gradient should be sufficiently long enough to facilitate proper separation.

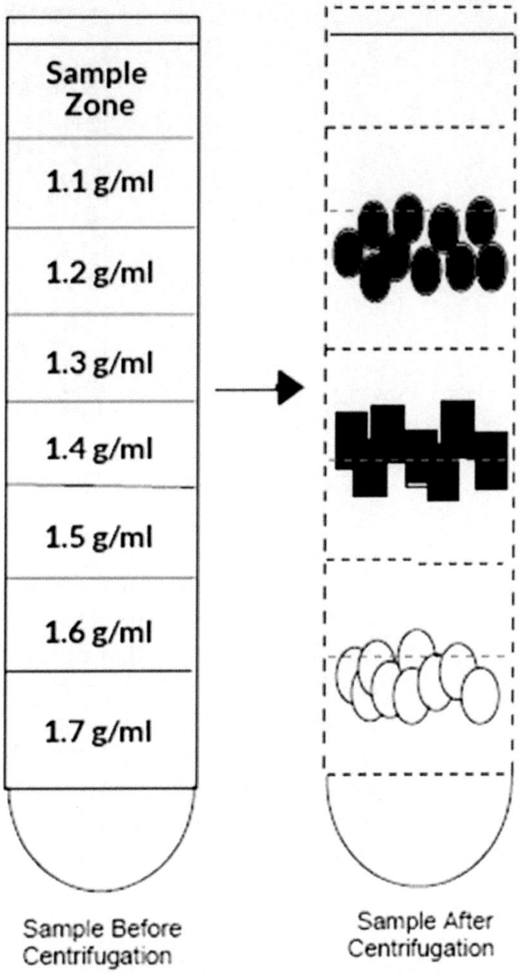

Fig. 10 Rate-zonal centrifugation [11]

Application of Rate-Zonal Density Gradient Centrifugation

It is documented that ribosome subunits and polysomes were the first subcellular particles to be fractionated using this technique. Separation of viruses is carried out using this technique as they differ in size and have unique density. RNA fractionation can also be accomplished by using sucrose gradient. It is also used for separating, purifying, and fractionating the DNA from both bacteria and viruses.

4.2.2 Isopycnic Centrifugation

Also referred to as equilibrium centrifugation, the particles get separated on basis of their densities rather than their size. The centrifuge tube is first filled with the gradient forming solution and the sample is loaded either on the top or bottom of the gradient or can be mixed in with the self-forming gradient. Thus, under the influence of centrifugal force, the sample and the gradient forming material, for example, cesium salt gets uniformly distributed in the centrifugal tube (Fig. 11). The cesium salt tends to redistribute

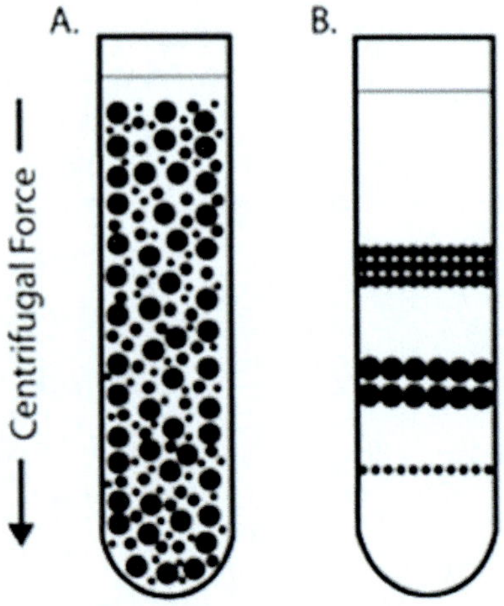

Fig. 11 Isopycnic centrifugation [12]

itself due to centrifugal force and forms a density gradient in the tube. The sample particles get separated to form bands when they reach the point or the region in the gradient where their buoyant density equals with that of the gradient. Thus the separated particles form an equilibrium with the salts as they get separated on basis of their densities rather than velocity. The particles keep on sinking through the gradient during the centrifugation till the density of the particle matches with that of the surrounding solution. The length of the centrifugation tube does not influence in attaining the equilibrium as the migration of the particle is dependent only on the density. For example, isolation of nucleic acid by CsCl gradient.

Applications

Large volumes of biomolecules can be purified using isopycnic centrifugation. Densities of various particles can be determined using this technique.

Table 2 compares the rate-zonal and isopycnic centrifugation.

4.3 Analytical Centrifugation

This technique is the combination of ultracentrifugation and optical monitoring system. An optical detection system is used to monitor the sedimentation profile of the sample in real time. UV and/or interference optical refractive index sensitive system is used to detect the sample. As a consequence of the applied centrifugal field the operator can track the evolution of sample concentration versus the axis of the rotational profile. Nowadays with the advancement in the instrumentation, the results are electronically digitized, recorded, and saved for mathematical analysis, when required. Figure 12 shows the design of an analytical centrifuge. These types of centrifuge operate at a centrifugation speed of 500,000 × *g*. They consist of an optical system, a refrigerated chamber with vacuum, rotor, and motor. Optical system has light

Table 2
Isopycnic vs. rate-zonal centrifugation

	Isopycnic	Rate-zonal centrifugation
Synonym	Sedimentation equilibrium, density equilibrium	Sedimentation velocity, S-zonal
Gradient	Steep Gradients applied can be discontinuous or continuous Maximum gradient density is adjusted such that it is higher than that of the most dense sedimenting particles	Shallow Gradient applied is continuous Maximum gradient density is adjusted such that it is less than that of the least dense sedimenting particle
Centrifugation	At equilibrium sedimentation achieved is complete Long time and high speed	Sedimentation is incomplete Short time and low speed
Separation	Peroxisomes, mitochondria, lysosomes, plasma lipoproteins, DNA, etc	Ribosomal subunits, DNA–RNA hybrids, etc

absorption system, two cells, that is, counterpoise and analytical cell, Rayleigh interferometric system, and Schlieren system.

This technique generates information like shape, confirmation, conformational changes occurring if any, stoichiometry, and the number of subunits that form the protein complexes. Real time monitoring of the samples during the centrifugation is carried out by the presence of the UV/Vis light-based optical scanning detection system. This lets you watch the sedimentation process and allows you to see the sample concentrate as the centrifugal force increases. Sedimentation equilibrium experiments and sedimentation velocity experiments are mostly carried out using analytical centrifuge. Experiments of sedimentation velocity generate information about the size, shape, and mass of the molecule as the ultracentrifuge, detector, and the computer record the sedimentation process over time, simultaneously the steady state equilibrium of the sample in the solution is studied by sedimentation equilibrium experiments. This type of analysis generates information about chemical and mass equilibrium constants.

4.3.1 Applications of Analytical Centrifugation Include

- Purity of macromolecules can be determined.
- Ligand-binding study.
- Change in relative molecular mass of super molecular complexes.
- Assessment of the conformational change in the structure of protein.
- Determination of the relative molecular mass of solute (within 5% SD).

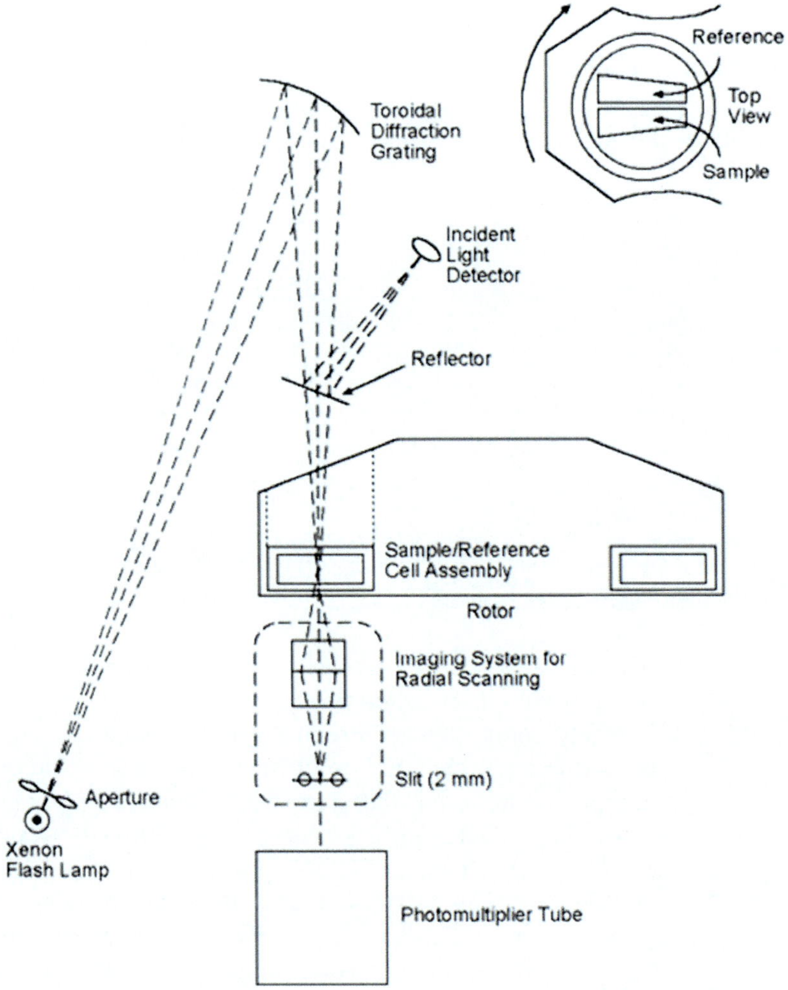

Fig. 12 Design of an analytical centrifuge [13]

5 Applications of Centrifuge

In clinical laboratory, centrifugation is used to

- To separate subcellular organelles like DNA and RNA.
- For removal of cellular elements from blood to obtain serum or cell-free plasma.
- For removal of chemically sedimented or precipitated protein from a sample.
- For analyzing the hydrodynamic properties of macromolecules.
- In immunochemical assay for separating the protein bound from the free ligand.

- Extraction of solutes present in the biological fluids.
- Separation of two miscible substances.
- Separation of lipid components.
- Freeing water from chalk powder.
- To produce skim milk.
- In cyclone separator for separation of finely dispersed particles in air.
- In forensic and research labs for separating blood and urine components.
- Separate proteins by means of purification techniques like salting out, for example, ammonium sulfate precipitation.
- Clarification and stabilization of wine.

6 Micellar Casein (MC) Preparation by Ultracentrifugation

Principle: Discussed in Subheading 1.

Materials:

1. Skim milk.
2. Whole milk.
3. Centrifuge tubes.
4. Ultracentrifuge.

Procedure:

1. Fill the centrifuge tubes in equal quantity with milk sample and apply the screw caps on the tubes.
2. Enter the centrifugation parameters of the ultracentrifuge, that is, time, vacuum, temperature, rcf/rpm.
3. For preparation of MC from whole or skim milk the ultracentrifuge should be operated at a speed of 30,000 rpm for 45 min.
4. Put off the vacuum after 45 min, and wait for the instrument to rest.
5. Observe the tubes for separation.

Observations

In case of whole milk, three layers will be observed, while skim milk gets separated into two layers. The pellet settled at the bottom is of MC (native), while the middle liquid layer consists of serum proteins and the top most layer contains milk fat in case of whole milk [14].

References

1. Christian GD (2007) Analytical chemistry. Wiley, Hoboken, NJ, USA
2. Nielsen SS (2017) Introduction to food analysis. Springer, New York, pp 3–16
3. Wilson K, Walker J (2010) Principles and techniques of biochemistry and molecular biology. Cambridge University Press, Cambridge
4. Pearson D (1976) The chemical analysis of foods. Longman Group Ltd, London
5. https://www.thomassci.com/Equipment/Centrifuges/_/Eppendorf-Refrigerated/Heated-Centrifuge-Model-5702RH
6. https://www.kindpng.com/imgv/TxxJhoJ_high-speed-centrifuge-hd-png-download/
7. https://www.h-saur.com/england/index.html
8. https://www.slideshare.net/HariSharanMakaju/pipettes-and-centrifuge-with-centrifugation
9. https://www.gmi-inc.com/product/himac-p100at2-fixed-angle-rotor%E3%80%80/
10. http://stevegallik.org/cellbiologyolm_fractionation.html
11. https://www.masterflex.com/tech-article/basics-of-centrifugation
12. https://www.sigmaaldrich.com/IN/en/technical-documents/technical-article/protein-biology/protein-pulldown/centrifugation-separations
13. https://www.nanolytics.de/en/analytical_ultracentrifugation/introduction
14. Mann B, Sangwan R, Kumar R (2008) Research techniques. In: Dairy chemistry. NDRI, Karnal

Chapter 4

Polyacrylamide Gel Electrophoresis

Abstract

Polyacrylamide Gel Electrophoresis is based on the principle of migration of charged particles under the influence of electric field to separate out proteins and nucleic acids. This chapter outlines this technique with respect to the separation of milk proteins along with the most frequently used protocols of gel media, buffer system, sample preparation, gel staining and visualization techniques. There are three commonly used PAGE techniques, namely, native PAGE (based on net charge, size, and shape of proteins), SDS PAGE (based on size of proteins), and urea PAGE (based on net negative charge and size of proteins) which can be utilized for various applications during analysis of milk proteins. Moreover, selection of correct visualization and detection techniques after separation of proteins over gels is as important as the protocol itself during electrophoresis.

Keywords Native PAGE, SDS PAGE, Urea PAGE, Casein, Whey protein, CBB staining

1 Introduction: Principle, Components, and Gel Media

Polyacrylamide gel electrophoresis, popularly known by its acronym PAGE is an analytical technique which is based on the principle of migration of charged particles under the influence of electrical field. The main purpose of this technique in analytical chemistry is to separate the mixture of protein or nucleic acid based on their size. The technique also provides fairly good idea about the molecular weight of the proteins. For this chapter, we will discuss this technique with respect to separation of proteins only. At the pH value other than the isoelectric point (p*I*), proteins possess either positive or negative charge and migrate under the influence of electric field, when a potential difference is applied across the electrodes. The force which moves these particles comes from the applied potential difference, which generates potential gradient. During migration, these molecules face the opposite frictional force, intensity of which depends on the type of matrix, porosity of the gel, buffer viscosity, hydrodynamic size, shape and net charge of the molecule. In general, smaller protein molecules migrate faster and vice versa. The electrophoretic mobility depends on the

Kamal Gandhi et al., *Advanced Analytical Techniques in Dairy Chemistry*, Springer Protocols Handbooks,
https://doi.org/10.1007/978-1-0716-1940-7_4,

applied voltage, net charge of the molecule, and friction according to the formula:

$$\text{Electrophoretic Mobility} = \frac{\text{Applied Voltage} \times \text{Net Charge}}{\text{Friction}}$$

According to this equation, under constant-voltage conditions, the electrophoretic mobility of a molecule depends only on the charge and frictional force. For similar particles, frictional force varies with size and not its shape. Therefore, the remaining variables are charge and size. When charge is taken care of (e.g. with negatively charged sodium dodecyl sulfate) the velocity of protein depends only on the charge-to-mass ratio.

Other factors affecting separation of proteins during electrophoresis are: (1) electroendosmosis, charged groups present on the surface of support media may hinder the movement of oppositely charged particles in the system and affect the overall flow, for example, silanol group and carboxyl group of glass and paper, respectively, (2) heat generation, constant voltage increases the current which further increases heat output in the system. Excessive heat can denature sensitive proteins leading to a decrease in viscosity [1].

Two types of geometries are available in different sizes for electrophoretic assembly, that is, vertical and horizontal. Further, two more arrangements of the gel matrix depending on the shape of gel are there: column and slab. Slab gels are more frequently used than column gels, because it is easier to make wells in slab gel wherein many samples can be loaded in different wells unlike in column gels where single sample is added on the top of the gel. Therefore, mini-vertical slab gel units (Fig. 1) are most popularly used for polyacrylamide gels across different laboratories of the world.

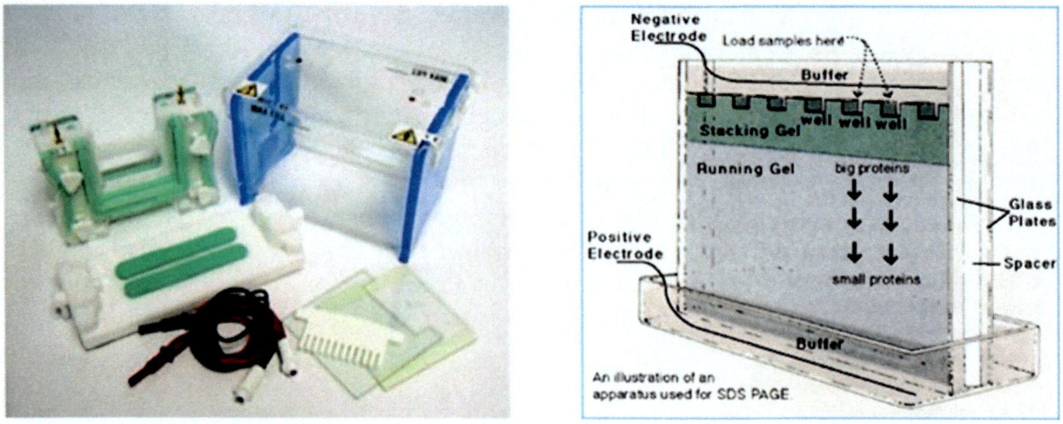

Fig. 1 Mini-vertical slab gel electrophoresis unit [2]

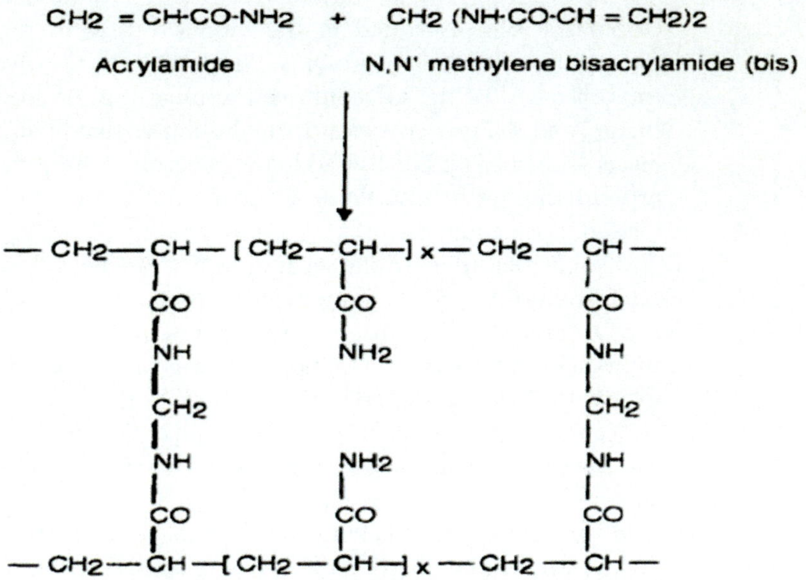

Fig. 2 Polymerization of acrylamide [1]

Other components of the equipment are glass plates which are clamped together after inserting spacers at both sides for gel casting. Combs are used to make wells in the gel. Rest of the accessories includes buffer tanks, electric power pack, electric cables, etc. After casting gels, apparatus is assembled so that buffer surrounds the gel plates. Samples are loaded and power pack supplies current between the electrodes in the electrophoretic assembly. Also, there is a list of support media which can be used as matrix which act as a porous mechanical support for protein separation. Cellulose, cellulose acetate, agarose, starch, and polyacrylamide (PA) crosslinked with the help of crosslinker (N,N′-methylene-bisacrylamide) are most popular (Fig. 2). Polymerization of acrylamide is a free radical reaction which is initiated by the addition of ammonium persulfate (APS) and catalyzed by N,N,N′N′-tetramethylenediamine (TEMED) which catalyzes the decomposition of APS to produce free radicals.

The most inconvenient part of this technique is gel casting. Although, in many laboratories, these gels are casted by the researchers, but commercially available precast polyacrylamide gels are quite popular these days. Apart from being handy, the use of precast gels is also comparatively safe as they eliminate the need of handling neurotoxic acrylamide monomer.

2 Sample Preparation and Buffer Systems

For milk protein separation, PAGE is usually carried out at alkaline pH, where most of the proteins are anionic and therefore move towards anode. The sample to be analyzed is loaded in the wells at the top of the gel and a "tracking dye" either bromophenol blue or Coomassie blue is added in the sample buffer to monitor the movement of colorless proteins. Movement of the dye band to the other end of the gel confirms the migration of protein across the gel and the power pack is turned off after that. Thickeners like glycerol or sucrose are also added in the sample buffer so that the protein samples may settle down at the bottom of the well while loading the sample.

To further improve the separation of proteins, gels of different ionic strengths and pH values can be used. Here two buffers are used to prepare different gels and the system is known as "discontinuous buffer system." The upper layer is known as the "stacking gel" with 2–4% acrylamide and pH 6.8. As the name suggests, it concentrates the sample into narrow bands prior to their entry into the lower gel. The sample buffer and the running buffer which surround the gel contain glycine. Under the influence of electric field, glycine enters stacking gel, but most of it remains in the zwitter ion form with no net charge and hence has low electrophoretic mobility. To maintain a constant current, proteins (still negatively charged) move towards anode in front of the glycinate ion but behind the fast-moving chloride ions. Therefore, proteins get sandwiched between chloride and glycinate ions giving rise to narrow stacks/bands. The lower layer is known as the "resolving gel" or "separating gel" and contains 6–16% acrylamide and pH 8–9. At this pH, all the ions encounter increase in pH and a decrease in pore size which increases the frictional force. Here the glycine mostly exists as an anion rather than zwitter ion and the relative movement of anions shifts to chloride > glycine > proteins. The different types of proteins present in the sample will have different charge-to-mass ratio, depending on which the mobility differs and in the process these proteins get separated into different bands.

3 Types of Gel Electrophoresis Commonly Used for Milk Protein Separation

Different types of PAGE techniques depend on different parameters like hydrodynamic size, net charge, shape, buffer systems, etc. to effect protein separation [3]. Several modifications of PAGE are now available which have increased its usefulness and applications. The most commonly used formats of SDS-PAGE for the separation of milk proteins can be classified (Fig. 3) as following:

3.1 Native PAGE

It is also known as alkaline or non-denaturing PAGE. In this type of PAGE, protein separation is achieved in their native form on the basis of their size, shape, and net charge. Neither denaturing (SDS) nor reducing agent [2-mercaptoethanol (2ME) or dithiothreitol (DTT)] is used. As the SDS is not there to denature protein, a uniform charge is not achieved and the protein retains its bioactivity

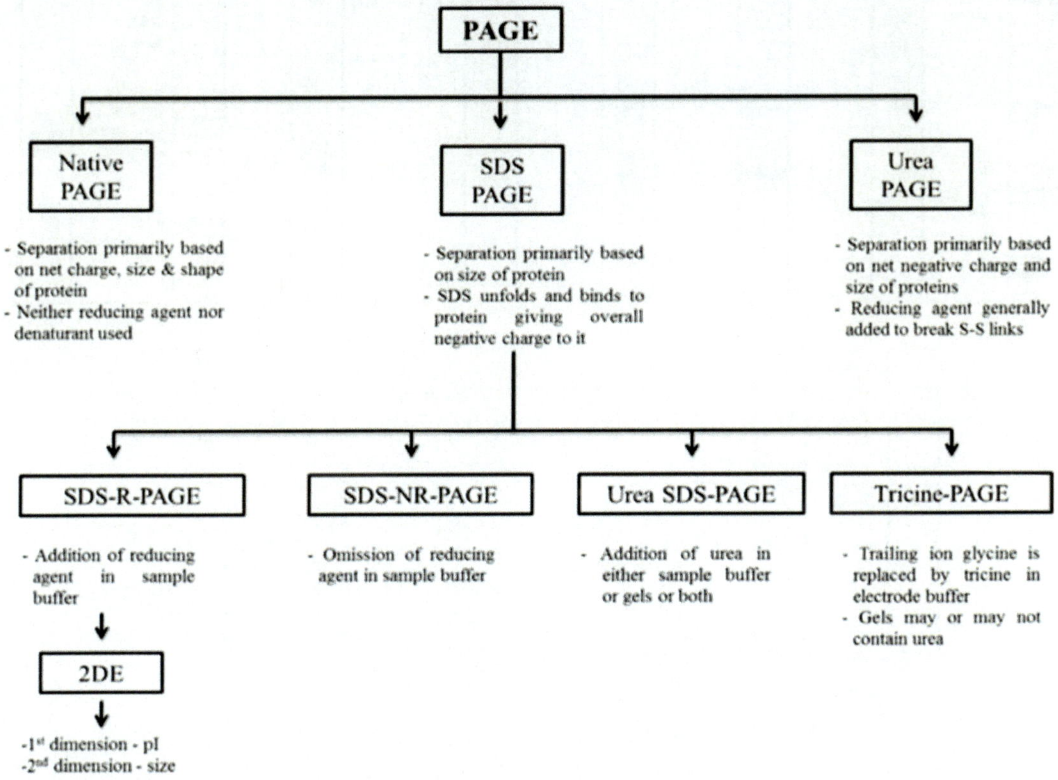

Fig. 3 PAGE techniques commonly used for milk protein separation [4]

even after migration on the gel. Also, in the absence of reducing agent, disulfide links will not break to bring proteins into linear polypeptide form and thus shape of the protein is retained which can have influence over the mobility. As far as milk proteins are concerned, native PAGE is suitable for the separation of whey proteins, but not caseins. The p*I* values of different casein fractions are very close to each other and fall between pH 4.9 and 5.8, making it difficult for them to get separated on the basis of charge and they migrate in close proximity to each other. Moreover, polymerization of κ-casein (with itself and other whey proteins) during processing of milk leads to the formation of higher molecular weight complexes which cannot be separated in the absence of reducing and denaturing agents. Conversely, whey proteins are compact globular molecules with discrete net charge and size, making it easier for them to migrate separately on gels. One of the limitation of native PAGE is that for the identification of proteins, purified standard proteins are run in the parallel lanes and commercially available molecular weight markers cannot be used for the estimation of size using it. Schematic diagrams and actual photographs (from literature) of milk protein separation on various types of PAGE techniques are shown in Fig. 4.

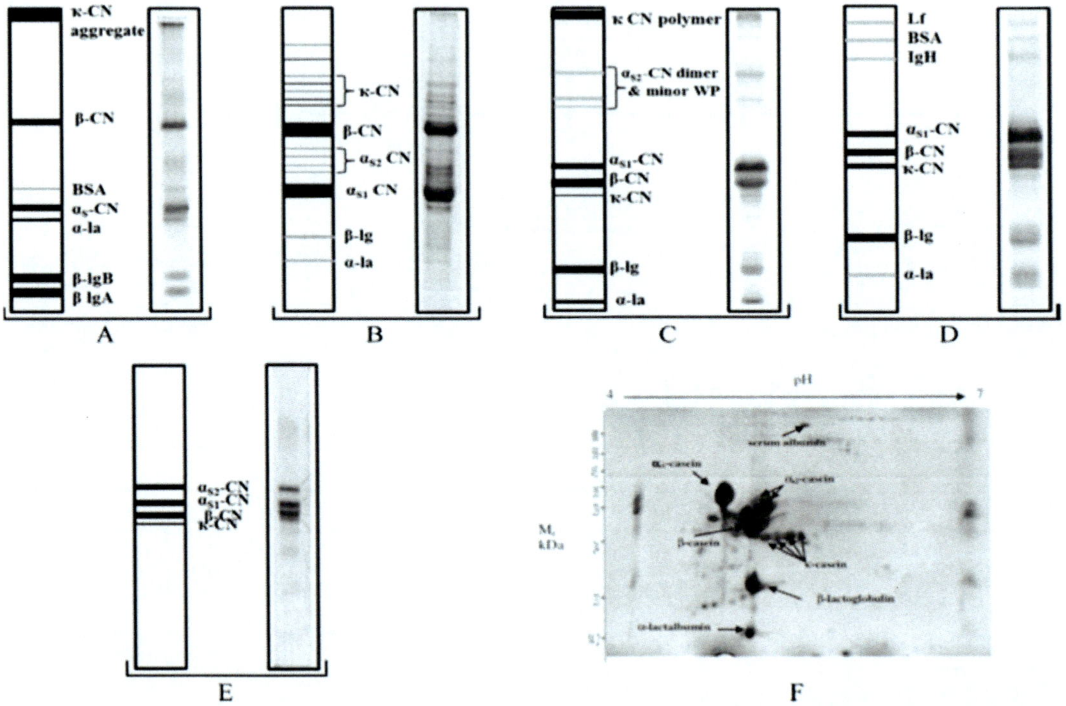

Fig. 4 Milk protein separation on various PAGE techniques [4]

3.2 Urea PAGE

For improving casein separation on PAGE, higher concentration of urea (up to 6 M) can be added to the gel buffer, sample buffer, and running buffer. Urea can denature and unfold the caseins and resolve them on the basis of differences in their net charge. Reducing agents are also added to dissociate disulfide bonds; otherwise, κ-casein and α_{s2}-CN (capable of making intermolecular disulfide links) do not separate properly under the conditions of urea PAGE. Conversely, whey proteins fail to separate properly on urea PAGE as discrete bands reduce its effectiveness for their separation. Unstability of urea in solution form can lead to changes in p*I* values of the proteins and thus diffused/smeared bands can appear after electrophoresis. Therefore, freshly prepared urea solutions should be used to avoid it.

3.3 Sodium Dodecyl Sulfate–Polyacrylamide Gel Electrophoresis (SDS PAGE)

Laemmli [5] is the most frequently used protocol for the separation of milk proteins. SDS (CH_3-$(CH_2)_{10}$-$CH_2OSO_3^{-}Na^{+}$) is an anionic detergent, used as a denaturant to bring compact or disordered protein molecules into linear rod like polypeptides. It is added in the samples buffer to which protein sample is added and heated for 3–5 min. The changes brought by SDS in the protein structure can bring tremendous impact on PAGE and can be enlisted as below.

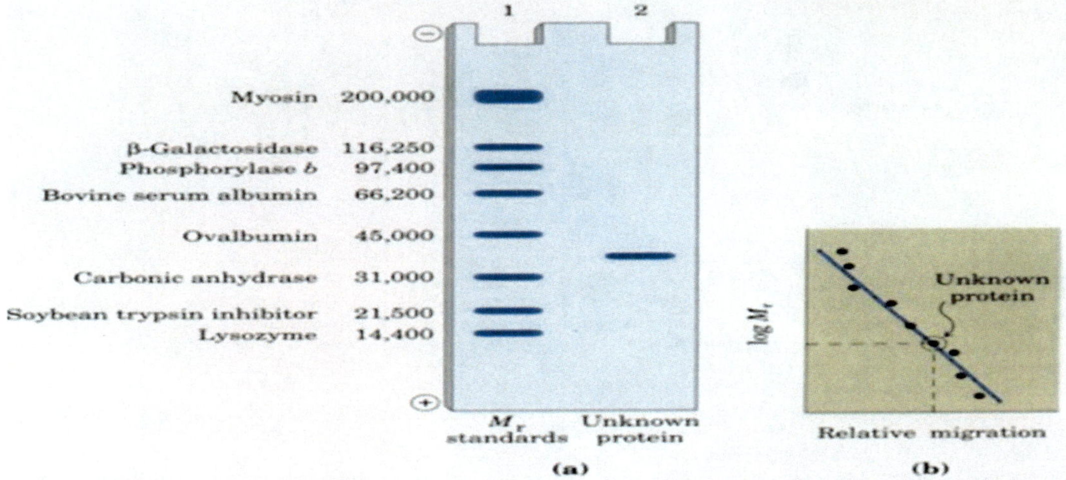

Fig. 5 Relationship between log molecular weight (M_r) and electrophoretic mobility or relative migration of proteins separated on gel [6]

- Effect on size of protein: On an average, one SDS molecule can bind to two amino acid residues. It binds to proteins, in general, at a constant ratio of 1.4 g per g protein. In principle, if the ratio of binding is constant, then the effect of extra size added by SDS on protein should get nullified in the results obtained.
- Effect on charge: Binding of SDS with protein masks the original charge of protein and produces a polypeptide chain of a constant charge-to-mass ratio.
- Effect on shape: Uniform rod like shape is produced (especially with the addition of reducing agents).
- Under these conditions of uniform charge/mass ratio and shape, electrophoretic mobility will depend primarily on the molecular weight of the protein as it migrates through the gel.
- Molecular size ladders of known proteins can be conveniently and reliably used as standards for the estimation of size of protein. The linear relationship between the electrophoretic mobility and molecular weight can be drawn on a graph as shown in Fig. 5. The electrophoretic mobility of proteins separated on the gel can be calculated with the help of the following formula:

$$\text{Electrophoreic mobility} = \frac{\text{Distance migrated by protein from start of separating gel}}{\text{Distance migrated by tracking dye from start of seprating gel}}$$

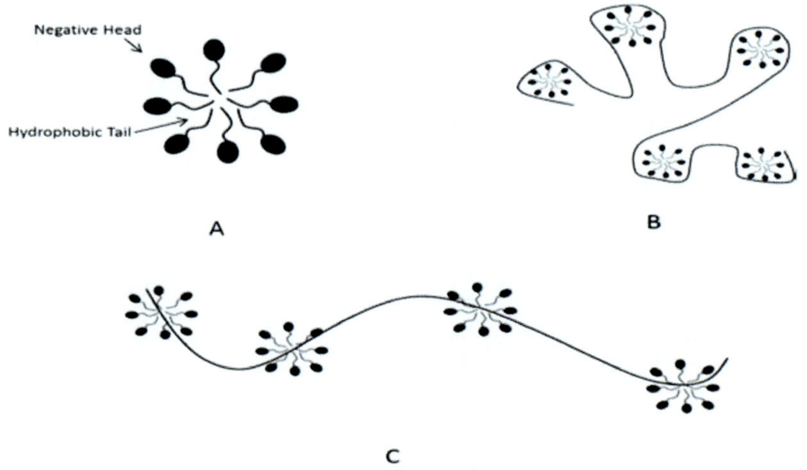

Fig. 6 Binding of SDS with bovine serum albumen (BSA) and casein. (**a**) SDS micelle, (**b**) interaction of SDS micelle with BSA, (**c**) interaction of SDS micelle with casein [4]

When only SDS is added into the sample buffer, without the addition of any reducing agent the PAGE is known as nonreducing SDS-PAGE and otherwise it is termed as reducing SDS-PAGE. As 2ME/DTT can reduce disulfide links, this converts the polymers of κ-CN, α_{s2}-CN, β-lg, etc. into monomers which leads to sharp bands corresponding to these proteins over the gels. This is especially required when processed milk samples are to be analyzed using SDS-PAGE.

Further, during SDS-PAGE of milk proteins, both caseins and whey proteins can be resolved simultaneously. While native PAGE and urea PAGE are recommended primarily for the separation of whey proteins and caseins, respectively, literature indicates that the reducing SDS-PAGE is popularly used for the simultaneous analysis of both caseins and whey proteins. Although SDS-PAGE can give superior results, for example, to analyze the changes brought by processing on milk proteins, but estimation of molecular size of caseins using SDS-PAGE can be misleading. It is because that SDS binding with caseins is not uniform and it binds differentially with caseins and whey proteins (Fig. 6). With whey proteins like bovine serum albumen (BSA) SDS molecule interacts via its negatively charged head, while with caseins it interacts via hydrophobic tails. Also, with caseins its interaction is not constant and may range from 0.9 to 3.4 g SDS per g of protein, leading to inconsistencies in charge-to-mass ratios. Also, as there is not much difference in the molecular weight of caseins, the bands often overlap in SDS-PAGE gels, if the concentration of milk sample in the wells is high.

3.4 Tricine PAGE

This technique is particularly useful for the electrophoretic separation of low molecular weight proteins and peptides (1–30 kDa). Here glycine is replaced by tricine in the buffer system. Although tricine SDS-PAGE can be done using two gels (stacking and separating), but in the original protocol the workers have recommended the use of three gels (stacking, spacer, and separating) with increasing concentration of acrylamide, which makes it somewhat more tedious than the other PAGE techniques. In this type of PAGE, when tricine is used as trailing ion (instead of glycine), it allows better resolution of smaller proteins (even 1 kDa) at lower acrylamide concentration and without addition of urea. Smaller proteins interact with SDS and form complexes of almost similar size as SDS micelles itself. Therefore, their separation from the bulk of SDS micelles is difficult. In the stacking gel, glycine and tricine behave quite differently when it comes to separation of smaller proteins. Glycine migrates very slowly and stacks very large proteins. Tricine, however, moves comparatively faster (despite its higher molecular weight) under the conditions of stacking gel, thereby shifting the stacking limit to lower molecular weight proteins/peptides.

3.5 Isoelectric Focusing (IEF)

Proteins are amphoteric/zwitter ion molecules. Below/above their p*I* values, they possess positive/negative charge and migrate towards cathode/anode. While at p*I*, they possess no net charge and thus do not move towards any electrode. On the basis of this very property of proteins, they can be separated over polyacrylamide gels (PAGs) and the technique is known as isoelectric focusing. The proteins having <0.02 pH unit difference in their p*I*s can be separated by IEF as this is one of the high-resolution protein separation techniques. PAGs of low acrylamide concentration (<4%) are generally used as separating matrix where proteins can move freely without any sieving effect/frictional force. In these gels, gradually changing pH gradient is established by adding ampholytes of different pH ranges. The chosen pH gradient can be broad (e.g. pH 2–10) or narrow (e.g. pH 6–7). The former will separate several proteins with wide range of p*I* values, while the latter is used for precise identification of p*I* of specific proteins.

In order to estimate pH gradient in the IEF strips, ampholyte (low molecular weight synthetic polyelectrolytes having large number of amino, carboxylic, and sulfonic groups) mixture is added into the gel mixture before polymerization. Once the gel is formed, these ampholytes are electrophoresed over the gel by applying potential difference across the two ends of the gel. Depending upon their place in the gel, these ampholytes either acquire positive or negative charge and move towards the opposite electrode. As they migrate up/down the pH gradient, they will either get increasingly protonated/deprotonated and their respective net charge will decrease towards zero. At some point during their migration, these ampholytes will no longer migrate and that point marks the p*I* of the respective ampholyte (Fig. 7). The protein sample is applied

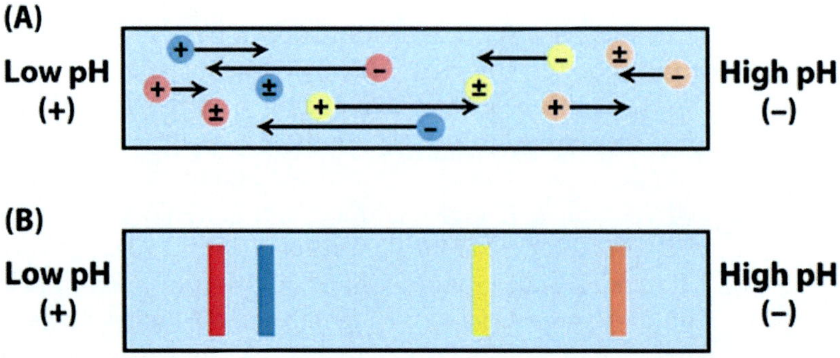

Fig. 7 Schematic representation of IEF strips before and after the establishment of pH gradient [7]

either in solution form over the gel (after establishing pH gradient) or can be added to the gel preparation directly. IEF can be used to determine microheterogeneity of proteins. For instance, a phosphorylated protein may appear as a single band in SDS-PAGE, but show several bands in IEF owing to its different phosphorylated forms (mono, di, and tri phosphorylated). It is because the addition of a few phosphate groups might not affect the size, but can influence the net charge and hence overall p*I* of the protein.

This technique can be used to identify most of the genetic variants of milk proteins (especially β-CN). On addition of urea and a reducing agent in the gel, the caseins separate in the decreasing order of their isoelectric point. The order of separation is γ-casein, κ-casein (A, B, C), α_{s2}-CN (A, D), β-CN (A1, A2, A3, B, C), α_{s1}-CN (A, B, C), while whey proteins separate in order of α-la B, β-lg (A, B, C) in the absence of urea. This technique can be applied routinely for milk phenotyping.

3.6 Two-Dimensional Gel Electrophoresis (2D-GE)

To separate a mixture of proteins, the IEF and SDS-PAGE can be used in combination and the resultant technique is known as 2DE (Fig. 8). In this technique the proteins are resolved by IEF in tube gels or strips. These strips with separated proteins are then placed over the SDS-PAGE vertical slab gel for further separation of proteins. It is estimated that more than 1000 proteins can be resolved by this technique. 2DE has been applied in dairy science for identification of genetic polymorphism, isoforms of different milk proteins, specific protein fractions in complex mixtures, milk of various species, and changes induced in the protein fractions of milk and milk products during processing and storage. Each species milk proteins resolved by this technique showed unique profile. Further the 2DE patterns of human, porcine, murine milk proteins were quite different from that of ruminant milk proteins.

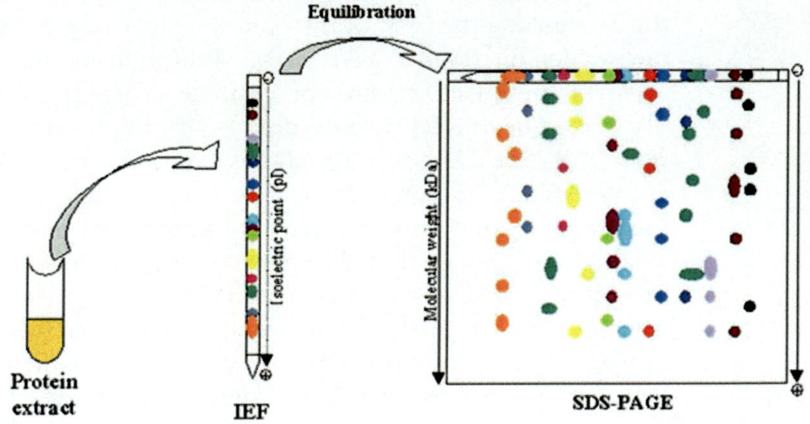

Fig. 8 Schematic diagram of two-dimensional gel electrophoresis [8]

4 Visualization and Detectiondetection

Identifying appropriate staining protocol is as important as selecting the correct type of PAGE for a particular application of milk protein separation. CBB is the most frequently used staining dye. It has detection limit of approximately 0.1–0.5 μg protein. It has two commonly used variants: CBB R 250 and CBB G 250 ("R" means reddish hue, "G" means greenish hue, and "250" is dye strength indicator). For milk protein staining, both CBB R 250 and CBB G 250 are used. Amido black is another anionic stain often used for staining proteins after PAGE. It can stain proteins within 10 min in a 1 mm thick gel, showing an obvious advantage over other dye used for visualization of milk proteins. The effectiveness of three stains, namely amido black, CBB R 250 and CBB G 250 has been compared using various staining protocols for cheese PAGE electrophoresis. They reported that overall amido black provides best results with cheese proteins, but CBB G 250 is better in staining the water-soluble nitrogen fraction of cheese (which did not stain with amido black) [9]. For decades these dyes have remained popular due to their reasonable sensitivity and easy protocol. But if more sensitivity is needed then metal-based stains can be used. Silver staining is used for the identification of minor milk proteins after separation on 2DE gels. It can give over hundred-fold increase in sensitivity over CBB. Despite being highly sensitive (50 pg per band), silver stain, like CBB, poorly detects highly negatively charged proteins. Therefore, cationic dyes like periodic acid schiff (PAS) stain and "stains-all" can be used. The kappa casein (κ-CN) and glycomacropeptide (GMP) can be stained using PAS staining. Also, sensitivity of GMP with "stains-all" is shown to be much more than that with CBB [10]. But the behavior of all negatively charged proteins present in milk is not similar with all cationic dyes.

After gel electrophoresis, densitometry is applied to quantify the proteins on the basis of the optical density of gels. After scanning of gel, the scanned gel is analyzed through appropriate image software for quantification, for example, TotalQuant, Image J, UVItec, QuantityOne, Advanced Quantifier, etc. Different researchers use 1D gel electrophoresis followed by densitometric analysis using densitometric image software. While scanning, the precaution should be taken that air bubbles and dust should be avoided and due care should be given to image processing transformation that can distort the information like change in brightness or saturation of image should be avoided. However, densitometry analysis of gels is not a preferred way to fully quantify because in gel electrophoresis smearing, streaking and wavy pattern of bands are commonly observed as staining intensity depends on the concentration of protein and particle density of protein in the band. Therefore, quantification of these type of bands may give unreliable and incorrect data interpretation and only semi-quantification of protein concentration should be done.

5 Chemistry of Milk Proteins Under Electrophoresis

Workers may come across many problems while resolving milk proteins over PAGE. There are more than hundred proteins/peptides in milk and milk products, all of which cannot be resolved efficiently using a single type of PAGE technique and therefore several modifications of PAGE have evolved to overcome these problems. Some of the merits and demerits of various types of SDS-PAGE techniques are compared in Table 1. Although broadly milk proteins can be divided into two major categories: casein and whey proteins, but several proteins with wide range of size, net charge and shape, having capability to make inter- and intramolecular interactions under commonly used milk processing conditions fall under these two categories. Understanding the chemistry behind the proteins and their behavior under specific conditions can help workers to identify the most effective PAGE technique among the available types to resolve them. Some of the specific examples are enumerated below.

- κ-CN makes intermolecular disulfide bonds with itself and other whey proteins. Therefore, native PAGE and nonreducing SDS-PAGE cannot give fine bands for its analysis. Conversely, reducing PAGE can be used for its analysis.
- Native PAGE resolves whey proteins best because of the compact structure of whey proteins as compared to the caseins. It can even resolve the beta-lactoglobulin (β-lg) into its two variants, that is, β-lgA and β-lgB.

Table 1
Merits and demerits of different types of SDS-PAGE techniques

Technique	Merits	Demerits
Native PAGE	• Protein remains undenatured and does not lose enzyme activity, if any • Distinct separation of whey proteins can be achieved	• Band intensity is relatively low • Molecular size cannot be estimated • Distinct separation of caseins cannot be achieved
Urea PAGE	• Better separation of caseins occurs, especially α_{s1}-CN and β-CN	• Molecular size cannot be estimated • Separation of κ-CN and α_{s2}-CN is not distinct
SDS-PAGE	• Relatively intense protein bands appear • Molecular size can be estimated	• Molecular size estimation for casein is misleading
Tricine PAGE	• Superior separation of low molecular weight proteins (1–30 kDa) • Four major casein bands can be separated. That is, α_{s1}, α_{s2}, β, and κ-CN	• More time consuming and tedious than other techniques • Smaller proteins resolve at the cost of larger ones
2DE	• Very high resolution of proteins can be achieved	• More time consuming and tedious than other techniques

- For β-CN and α_{s1}-CN separation over urea PAGE may give best results, but κ-CN and α_{s1}-CN appear as numerous light bands. Moreover, whey proteins cannot be resolved properly over them.
- Caseins have molecular weight in close proximity to each other. So, to resolve them using SDS-PAGE, low sample concentration should be used to avoid overlapping of casein bands.
- Using tricine SDS-PAGE, sharpness of casein bands can be achieved. Also, smaller peptides formed by processing of milk can be resolved in it properly.

6 Applications in Dairy Science

PAGE is often used to determine the protein composition of milk and milk products. Some of the specific applications need to be mentioned. For example, the study of effect of processing (heat, high pressure, pH, concentration, enzyme action) treatment on milk can be studied using these techniques. Moreover, identification or determination of different protein variants (e.g. A1 and A2 β-CN variants) can be achieved using PAGE. Apart from this, interspecies adulteration detection (e.g. addition of bovine milk in caprine milk) has also been studied by many workers.

7 Tricine-SDS-PAGE Protocol

Modified tricine-SDS-PAGE on 16% acrylamide gel as described by Schagger [11] is described below.

7.1 Reagents

1. Acrylamide solution (49.5% T, 3%C): Dissolve 48 g acrylamide and 1.5 g bisacrylamide in 40 ml distilled water and make the volume to 100 ml with the same. Filter the solution through Whatman No. 1 filter paper and store at 4 °C in amber-colored glass bottle.
2. Gel buffer (3×): Dissolve 36.6 g Tris and 0.3 g SDS in 60 ml distilled water. Adjust the pH to 8.45 with concentrated HCl and make the final volume to 100 ml with distilled water. Store the buffer at 4 °C.
3. Cathode buffer (1×): Dissolve 12.11 g Tris, 17.92 g Tricine, and 1 g SDS in distilled water and make the final volume to 1 l with distilled water. The pH of the solution should be approximately 8.2. Do not correct the pH.
4. Anode buffer: Dissolve 6.05 g Tris dissolved in 50 ml distilled water. Adjust the pH to 8.9 with concentrated HCl/1N HCl and then make the volume up to 500 ml with distilled water.
5. Sample buffer (4×): Prepare 4× sample buffer by mixing the following solution and then make the volume to 20 ml with distilled water.
   (a) SDS (12%, w/v)—12 ml of 20% SDS.
   (b) Glycerol (30%, w/v)—6 g.
   (c) 2ME (6%, v/v)—1.2 ml.
   (d) CBB (0.05%)—0.0 1 g.
   (e) Tris/HCl (pH 7) (150 mM)—3 ml of 1 M Tris-HCl.
6. Marker (2.5 μl/well): Add 3.5 μl of molecular weight marker to 10 μl sample buffer (1.33×). For the preparation of 1.33× sample buffer, dilute 1 ml of 4× sample buffer with 2 ml distilled water.
7. 10% Ammonium persulfate (APS): Dissolve 100 mg APS in 1.0 ml distilled water. This solution is always prepared fresh.
8. TEMED: *N,N,N′,N′*-Tetramethylethylenediamine.
9. Tris/HCl (pH 7) (1 M): Dissolve 12.1 g Tris in 80 ml distilled water. Adjust the pH of the solution to 7 with concentrated HCl. Make up the final volume up to 100 ml with distilled water.
10. SDS stock solution (20%): Dissolve 10 g SDS in 50 ml water. Warm gently for complete dissolution.

Table 2
Materials required for the preparation of stacking and separating gel

Contents	Stacking gel (4%, 10 ml)	Separating gel (16%, 15 ml)
Distilled water	8.0 ml	3.5 ml
Acrylamide/bisacrylamide	1.0 ml	5.0 ml
Gel buffer	3.0 ml	5.0 ml
Glycerol	–	1.5 ml
10% APS	60 μl	30 μl
TEMED	10 μl	10 μl

11. Separating and stacking gel solution: For the preparation of stacking and separating gels, mix the contents as mentioned in Table 2.

7.2 Gel Preparation

Clean gel plates of the size 8 × 10 cm are assembled on casting stand of vertical slab gel unit using one 1 mm spacers. Then separating gel mix is poured down to the level of 2 cm from top into the gel plates. Distilled water is layered over the separating gel gently and the gel is allowed to stand until polymerization is complete. After polymerization, water is removed and stacking gel layered over the polymerized separating gel and a slot former (comb having 6 wells of 9 mm) is placed in position to allow polymerize. The slot former is removed and the gel plates are transferred to the gel unit filled with electrode buffer.

7.3 Sample Preparation

Forty-five microliter of sample containing 25–45 μg protein is mixed with 15 μl of 4× sample buffer in a microcentrifuge vial. The buffer containing samples are heated on boiling water bath for 2–3 min (using floaters) and then allowed to cool to room temperature.

7.4 Electrophoresis

The gels are pre-run at 50 V for 30 min and then samples are applied (10 μl). After application of the samples, the electrophoresis is run at a constant voltage of 50 V through the stacking and at 90 V through the separating gel until the tracking dye front is closed to the bottom of the gel slab. The temperature during electrophoresis run should be kept at 8 °C by placing the whole electrophoresis unit inside the cooling cabinet.

7.5 Gel Staining

7.5.1 CBB

Reagents

1. Fixing solution: 10% trichloroacetic acid.
2. Staining solution: 0.025% CBB in 10% acetic acid.
3. Destaining solution: 10% acetic acid.

Procedure

At the end of the electrophoresis, the gels are removed from the electrophoresis unit and kept in fixing solution for 30 min followed by immersing the gel in staining solution. After staining for 1 h, the gels are transferred to the destaining solution at room temperature. Destaining is done, till the blue protein bands appear and background becomes clear.

7.5.2 Silver Staining

The protein bands can be visualized by adapting the silver staining method of Blum et al. [12] and the protocol is given below.

Reagents

All the solutions to be used in silver staining should be made fresh.

1. Solution A: 50 ml methanol is mixed with 5 ml acetic acid and volume made up to 100 ml with distilled water.
2. Solution B: 50% Methanol.
3. Solution C: Dissolve 0.2 g sodium thiosulfate ($Na_2S_2O_3.5H_2O$) in distilled water and make up the volume to 1 L.
4. Solution D: 0.2 g silver nitrate ($AgNO_3$) is dissolved in distilled water and volume is made up to 100 ml. The solution is chilled to 4 °C before use.
5. Solution E: 3 g of anhydrous sodium carbonate (Na_2CO_3) is dissolved in distilled water and volume is made up to 100 ml. Then, 25 μl of 37% formaldehyde solution is added to it and the mix the solution properly.
6. Solution F: 1.4 g Na_2-EDTA (ethylenediamine tetra acetic acid disodium) is dissolved in distilled water and volume is made up to 100 ml.

Procedure

The various steps of silver staining are shown below (Table 3):

Dark brown colored bands of proteins appear on the gel, against a light yellow colored background gel after staining. If development period is allowed to proceed for too long, dense protein zones will become saturated and negative staining will occur, leading to serious problems. In such case, gel is destained and restained properly. For the destaining, gel is dissolved in 0.4 g potassium ferricyanide ($K_3Fe(CN)_6$) in 200 ml of solution C. Destaining is done until no bands are visible; the gel had a yellow hue. Then the gel is washed 4–5 times for 15 min with milli-Q H_2O until gel become transparent and showed no background color. The gel was stained again starting with solution C of the gel staining protocol.

Table 3
Steps for the preparation of silver staining

Sr. No.	Solution	Operation	Time
1	Solution A	Fix	30 min
2	Solution B	Incubate	15 min
3	Milli-Q water	Wash 5 times	5 × 5 min
4	Solution C	Incubate	60 s
5	Milli-Q water	Wash 2 times	2 × 60 s
6	Solution D	Incubate	25 min
7	Milli-Q water	Wash 2 times	2 × 60 s
8	Solution E	Develop	10 min
9	Solution F	Stop develop	10 min
10	Milli-Q water	Wash 2 times	2 × 60 s

7.5.3 PAS Staining

Reagents

1. Schiff reagent: 2.5 g basic fuchsin is dissolved in 450 ml of boiling distilled water. It is cooled to 50 °C and 50 ml of 1 N HCl was then slowly added to it. Again, it was cooled to 25 °C and 5 g potassium metabisulfite ($K_2S_2O_5$) was dissolved into it. After shaking for 3 min, the solution was incubated in dark at room temperature for 24 h. To it, 5 g of fine activated charcoal is added and shaken for 3 min. The solution is filtered with Whatman No. 1 and stored at 4 °C in a foil covered bottle.
2. 50% Methanol.
3. 2% w/v Periodate solution: 2 g sodium periodate was dissolved in distilled water and volume made up to 100 ml.
4. 2% w/v Sodium metabisulfite solution: 2 g sodium metabisulfite is dissolved in distilled water and volume made up to 100 ml.

Procedure

The method of Jay et al. [13] can be followed for staining of the gels after SDS-PAGE. Gel is fixed for overnight in 50% (v/v) methanol solution with gentle agitation. It is allowed to swell in milli-Q water for 20 min. The water is replaced with periodate solution for 15 min. Gel is washed with several changes of milli-Q water for 2 min and treated with the concentrated Schiff's reagent. The gel is gently agitated in a fume hood until the bands turned magenta. The gel is again washed with milli-Q water and reduced with sodium metabisulfite for overnight. Gel was washed for 30 min with several changes of milli-Q water until the water is clear.

References

1. Walker JM (2006) Electrophoretic techniques. In: Wilson K, Walker J (eds) Principles and techniques of biochemistry and molecular biology. Cambridge University Press, New York
2. Abdallah E (2009) Antibacterial activity and toxicological studies on the oleo-gum resins of Commiphora molmol and Boswellia papyrifera
3. Mann B, Sangwan R, Kumar R (2008) Research techniques in dairy chemistry. NDRI, Karnal
4. Sharma N, Sharma R, Rajput YS, Mann B, Singh R, Gandhi K (2020) Separation methods for milk proteins on polyacrylamide gel electrophoresis: critical analysis and options for better resolution. Int Dairy J 114:104920
5. Laemmli UK (1970) Cleavage of structural proteins during the assembly of the head of bacteriophage T_4. Nature 227L:680–685
6. http://www.bioinfo.org.cn/book/biochemistry/chapt06/bio2.htm
7. https://slideplayer.com/slide/10339253/
8. https://www.creative-proteomics.com/blog/index.php/two-dimensional-gel-electrophoresis-2-de/
9. Creamer LK (1991) Electrophoresis of cheese. Bull Int Dairy Fed 261:14e28
10. Sharma N, Sharma R, Rajput YS, Mann B, Gandhi K (2021) Distinction between glycomacropeptide and b-lactoglobulin with ‘stains all’ dye on tricine SDS-PAGE gels. Food Chem 340:Article 127923
11. Schägger H (2006) Tricine SDS-PAGE. Nat Protoc 1:16e22
12. Blum H, Beier H, Gross HJ (1987) Improved silver staining of plant proteins, RNA and DNA in polyacrylamide gels. Electrophoresis 8:93–99
13. Jay GD, Culp DJ, Jahnke MR (1990) Silver staining of extensively glycosylated proteins on sodium dodecyl sulfate-polyacrylamide gels: enhancement by carbohydrate-binding dyes. Anal Biochem 185:324–330

Check for updates

Chapter 5

Western Blotting

Abstract

Western blotting is used to transfer proteins from polyacrylamide gels (PAGs) to the membranes, making its identification and determination more convenient. For this, PAG and membrane are sandwiched together between two parallel electrodes. On passage of current, proteins transfer from the gels to the membranes. This chapter outlines various components, merits and demerits of western blotting technique. Most commonly used blotting membranes, namely, nitrocellulose membranes, polyvinylidene difluoride (PVDF), and nylon, are also compared for their properties, merits and demerits in this section. Further, visualization and detection of protein markers and targeted proteins over the membrane can be achieved using various protocols. Ponceau S or Congo red dyes are frequently employed for the detection of protein markers, whereas for the detection of analyte (specific protein) antigen antibody interaction, enzyme-based interaction of carbohydrate moiety reaction can be utilized. Lastly, its applications in dairy science are also outlined.

Keywords Blotting, Nitrocellulose, Polyvinylidene difluoride, Nylon, Immunoassay

1 Introduction: Principle and Components

Western blotting is referred to as protein blotting technique wherein polyacrylamide gel (PAG), onto which proteins have been electrophoresed, is cossetted/sandwiched together with the microporous membrane. The cassette is then immersed in a buffer, placed between two parallel electrodes and current is passed at right angle to the gel cassette. This causes the proteins to electrophorese from the gel onto the membrane. Once transferred, the proteins can be analyzed on the membrane further for identification and determination (Fig. 1). The advantages and disadvantages of western blotting technique are compared in Table 1. To overcome the problems (as mentioned in Table 1) related to this technique some modifications have been suggested. For instance, efficiency of transfer can be increased with the help of heat, partial digestion of proteins before transfer and use of special buffer systems [2, 3]. Some of the important terminologies related to the technique are as follows:

Kamal Gandhi et al., *Advanced Analytical Techniques in Dairy Chemistry*, Springer Protocols Handbooks,
https://doi.org/10.1007/978-1-0716-1940-7_5,

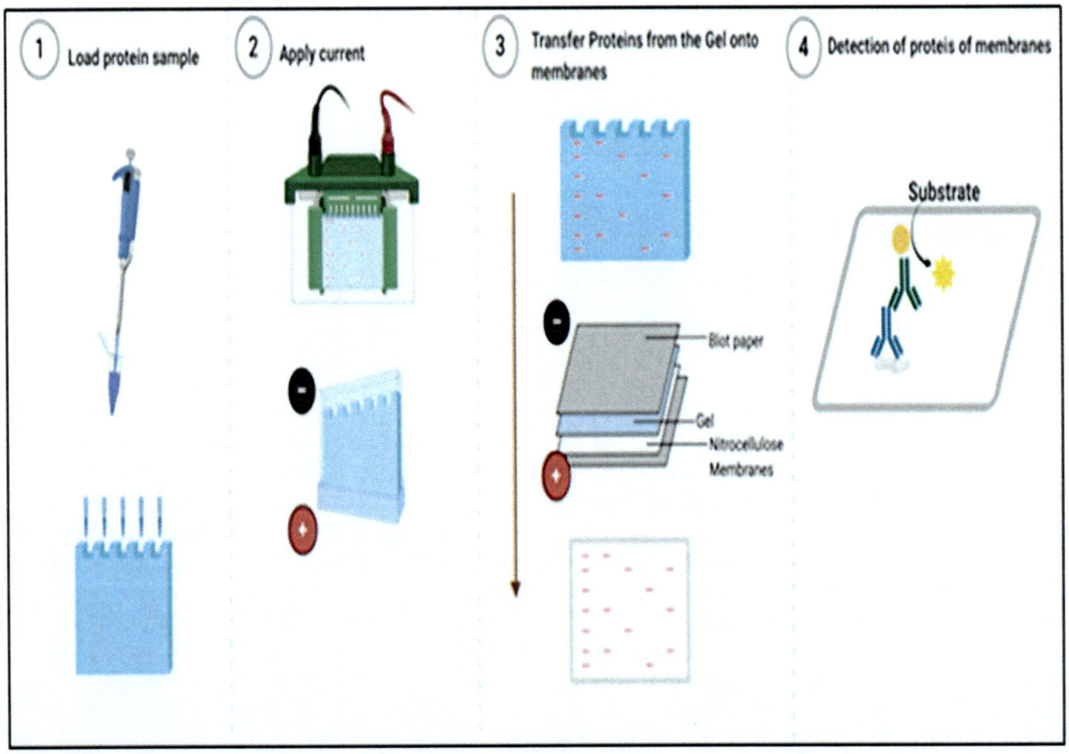

Fig. 1 Diagram representing the outline of western blotting [1]

Table 1
Advantages and disadvantages of western blotting technique

Advantages	Disadvantages
Strength: The wet membranes are pliable and strong enough to wash with different kinds of buffers repeatedly unlike soft gels	*Process time*: Overall long process time
Accessibility: As proteins bind to the surface of the membrane, their accessibility to the ligands increases as compared to the porous membranes where proteins are trapped inside the gel	*Efficiency*: As many steps are involved, sensitivity/accuracy of the results depends on the efficacy of many steps like transfer of proteins, protein retention over membrane, and final detection process
Sensitivity: Even very low concentration of proteins can be identified after western blotting	*Higher molecular weight proteins* blot poorly from the gels onto the membranes
Materials: Only limited reagents and other materials are required	*Technology intensive*: Requirement of additional blotting equipment. A good technical knowledge and hands on experience is required to get the good quality results
Storage: Prolonged storage of transferred proteins is possible	
Reusability: Same blot can be used for multiple analyses	

- Blot: Microporous membrane after transfer of proteins is known as blot.
- Southern blotting: It is the technique of transfer of DNA from gels to the membrane. The technique is named after the biologist (Edwin Southern) who first suggested it. Then came the RNA (or northern blotting) and protein (or western blotting) to respect the tradition of "geographic" naming in the honor of Edwin Southern.
- Dot blotting: Direct application of protein solutions onto the membrane rather than transferring from the gel.

2 Components of Western Blotting

The principal components of western blot technique include polyacrylamide gel with separated proteins, microporous membranes, the blotting equipment, and buffers. Selection of suitable membrane for a particular application is a crucial step. Some of the properties to be taken care of while selecting any membrane depend on specific properties like high volume-to-surface area ratio, high binding capacity, reproducibility of protein binding, pliability in various buffer systems, and long-term storage ability. The most commonly used membranes are cellulose, cellulose acetate, nitrocellulose, polyvinylidene difluoride or polyvinylidene fluoride (PVDF), nylon, and polyethersulfone. The properties, merits, and demerits of the commonly used microporous membranes are highlighted in Table 2.

Further, among the buffers, the Towbin buffer system is most commonly used which comprises tris (25 mM), glycine (192 mM), methanol (20%), SDS (0.001%). The methanol has important role here as it decreases pore size, inhibits swelling of gel, removes SDS from proteins, increases the hydrophobic interactions between membrane and proteins, and helps in the separation of low molecular weight proteins. It should be noted that for high molecular weight proteins, addition of methanol might cause problem as it can decrease their elution efficiency by retarding the elution from gel and also it can denature large proteins. SDS (up to 0.01–0.02%), however, aids in transfer of higher molecular weight proteins, especially after isoelectric focusing (when proteins do not have any net charge). However, it can decrease the binding of proteins to nitrocellulose and PVDF membranes. Another buffer known as CAPS buffer is used when SDS and glycine are not required. It contains 3-(cyclohexylamino)-1-propanesulfonic acid (10 mM), methanol (10%) (pH 11.0). It is used when membrane is to be used for protein analysis or sequencing (as it does not contain glycine).

Table 2
Comparison of commonly used blotting membranes

Membranes	Properties	Merits	Demerits
Nitrocellulose	• Cellulose is treated with nitric acid to replace -OH on each sugar unit by nitrate group • Hydrophobic interactions between membrane and protein	• Bound antigens can be removed using nonionic detergents • These membranes can distinguish between complexed and uncomplexed molecules • Proteins bound on these membranes can be stained with CBB (high background), amido black (high sensitivity), Ponceau S (low background)	• Fragile so difficult to use multiple times • Become brittle after drying
PVDF	• Linear polymer with repeating –(CF_2-CH_2)- subunits • Combination of dipole and hydrophobic interactions bind to proteins	• High protein binding capacity • Good mechanical strength • Good chemical stability with wide range of solvents • Long storage stability after blotting • Blot can be used for N-terminal sequencing and proteolysis • Staining can be done with CBB, silver nitrate, India ink, and amido black	• Pre-wet step is required with methanol or ethanol before using aqueous buffers
Nylon	• These have a matrix of polyhexamethylene adipamide which is modified by the addition of numerous tertiary amino groups • Used more for southern blotting rather than western blotting	• Excellent mechanical strength • High binding capacity compared to nitrocellulose membrane • More consistent and reproducible transfer results • High sensitivity	• Higher non-specific binding • Binds strongly to anionic dyes and therefore gives high background • Distaining is difficult unlike nitrocellulose membranes

Further, blotting can be done in two ways: semidry transfer (Fig. 2) and tank buffer transfer. In the former way, a stack of wetted filter papers surrounds the gel and the membrane, instead of liquid buffer in the reservoir tank. Also, stainless steel, graphite, or conducting polymer plates are used as electrodes to provide homogenous electric field. Some of the merits of semidry transfer include short run time (30–60 min), ease of handling, low buffer consumption, and negligible heat production when electric field is applied. But, fairly high current passes through the gel because of large cross-sectional area.

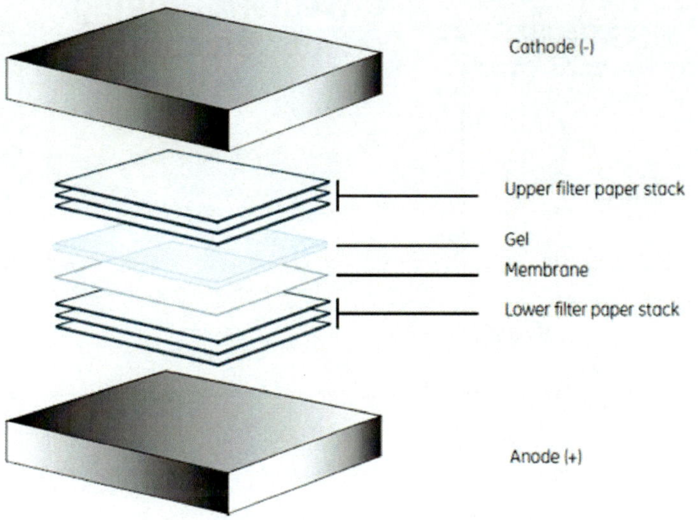

Fig. 2 Semidry transfer of proteins from gel to the membranes [4]

3 Visualization and Detection

3.1 Visualization of Protein Markers

Like SDS-PAGE, the proteins in the sample can be compared for molecular size estimation using protein standard/marker (containing a mixture of known proteins) which is run in parallel lanes and are blotted in the membranes. Generally, the lane containing protein standard is cut from the membrane and stained with dyes like congo red or Ponceau S (for nonspecific reaction of dye with protein) and the portion of blot containing sample is used for specific analysis, for example, antigen–antibody reaction using primary antibody and secondary antibody–enzyme conjugates. For staining of markers, generally stock Ponceau S (200 mg in 3% trichloro acetic acid) is used after tenfold dilution with distilled water. Destaining of membrane is done by rinsing it in distilled water or phosphate buffer saline (PBS) until clear pink bands over white background are observed. Ponceau S stains reversibly with proteins. Similarly, stock congo red dye solution (1% in distilled water) can be used after tenfold dilution with 0.2 M acetate buffer, pH 3.5. Destaining is done with distilled water until brown protein bands are observed against light pink background.

3.2 Visualization of Proteins in the Samples

Western blotting refers to the identification of specific proteins after its transfer to the membrane based on either molecular weight, antigen–antibody interaction, or reaction of carbohydrate moiety. After transfer from the gel to the membrane, the proteins retain the same pattern of separation they had on the gel. The blot (membrane with transferred protein) is then incubated with a generic protein (such as milk protein) to block rest of the sticky places on

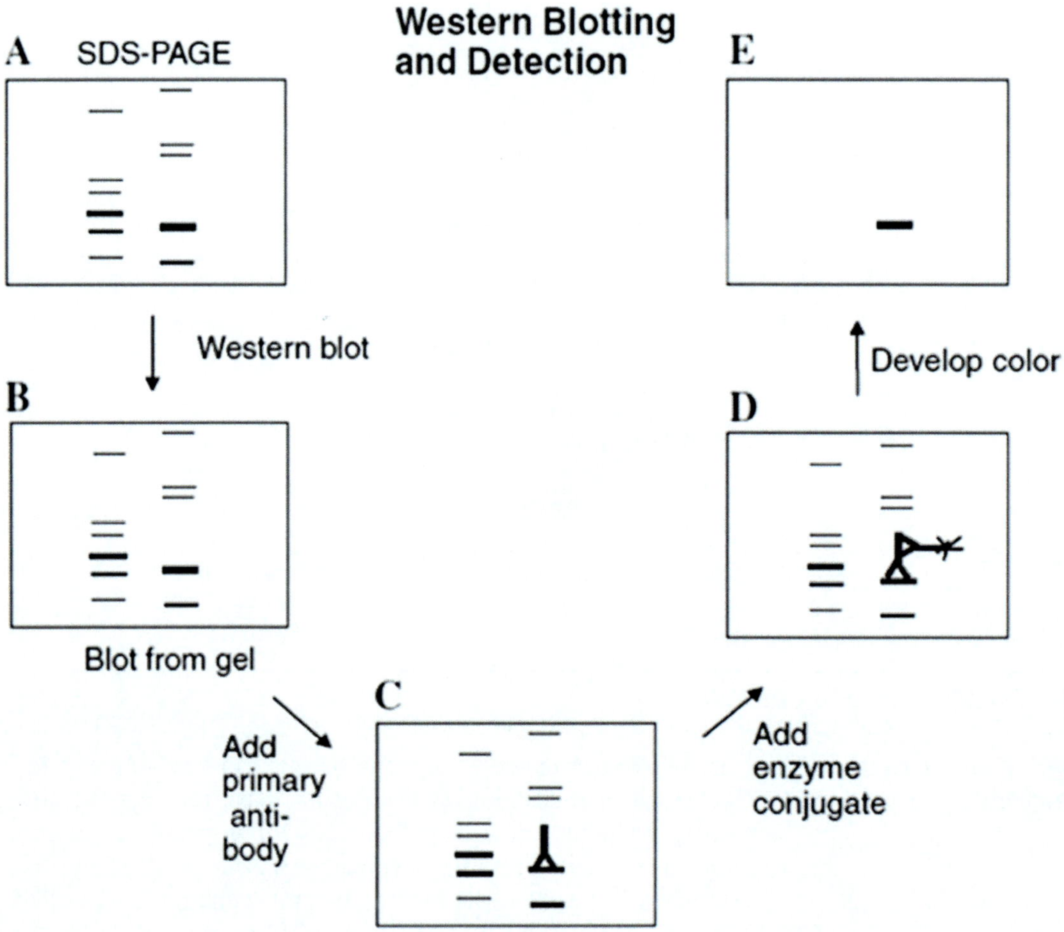

Fig. 3 Schematic representation of western blotting by the use of secondary antibody peroxidase conjugate mechanism [5]

the membrane. An antibody is then introduced into the system which can now bind to the antigen (present in the sample). The antibody is associated with an enzyme (e.g. alkaline phosphatase or horseradish peroxidase) or a dye, the color of which cannot be seen/revealed at this point. The location of the antibody can be revealed only after incubating the blot with a colorless substrate that attaches the enzyme or dye to give a colored product. The important steps under this are represented in Fig. 3.

The main steps after transfer of proteins on membrane include antibody probing, detection, and analysis. The detailed description of these steps in given below.

For the detection of antigen, first a non-labeled primary antibody (developed against the target protein or antigen) is made to bind to the antigen. Then a labeled secondary antibody (species specific) binds to the specific regions of primary antibody. Most of

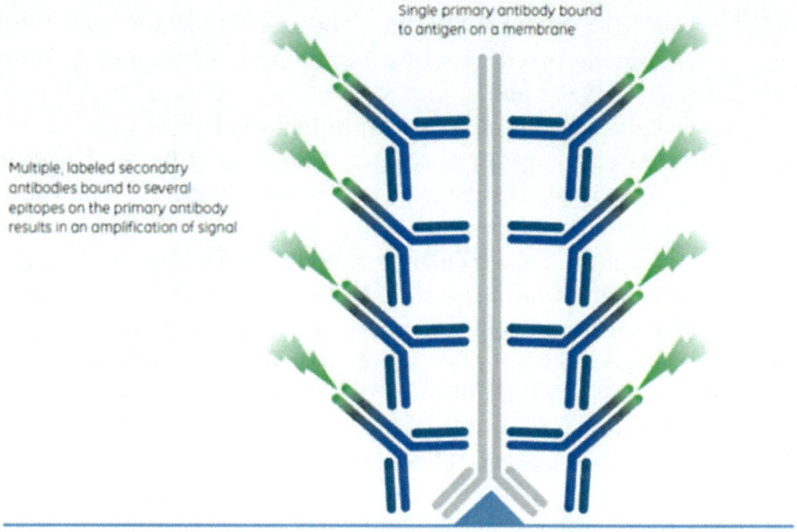

Fig. 4 Monoclonal non-labeled primary antibody reaction with polyclonal labeled secondary antibodies during western blotting [2]

the times, the secondary antibodies which are used for this purpose are polyclonal and therefore many of them can bind to the primary antibody which helps in the amplification of the signal that can be measured and is proportional to the concentration of antigen present over the blot (Fig. 4).

In order to have the abovementioned reaction to occur with sensitivity and specificity over the blot, it is essential to block any nonspecific binding of the antibodies over the membrane. There are many chemicals/agents available to block the background of the blot which can be categorized into two important classes, that is, proteins and nonionic detergents. Some of the examples of protein based blocking agents include bovine serum albumen (BSA), nonfat milk, casein, fish gelatin, serum, etc. Proteins generally attach permanently to the membranes via hydrophobic interactions, so proteins are often termed as permanent blocking agents as they have more affinity towards the membranes than the antibodies which are used in the subsequent washing steps. One of the important characteristics of blocking agent is that it should fill all the unoccupied sticky parts of the membrane without interfering in the antigen–antibody reaction. The choice of blocking agent and its concentration to be used in the blocking solution totally depends on the antigen–antibody pair used in the assay. There is not any universal blocking agent which can be used for all types of membranes without compromise and therefore optimization of such parameters is key to the success of western blotting assay. Conversely, detergents are termed as nonpermanent blocking agents as they do not get permanently attached to the membrane and can be removed by simple washing steps, for example, Tween 20, Triton X-100, SDS, NP40, etc. These are generally used together with protein based blocking agents and not alone. Commonly used buffers for preparation of blocking solutions are phosphate buffer saline (PBS) and Tris-HCl buffer saline (TBS). In assays, where the antigen is phosphorylated, TBS should be used instead of PBS; otherwise, primary antibodies may compete for phosphate present in the antigen as well as buffer.

Labeled secondary antibody can be chosen depending on the western blotting system to be developed. For instance, enzymes, biotinylated antibodies, gold conjugated antibodies, radioisotopes, fluorophores etc. are some of the options. As mentioned before, the two most common enzymes used for the detection are alkaline phosphatase and horseradish peroxidase (HRP). Both can be used for chromogenic, chemiluminescent, or chemifluorescent detection methods. The advantage of alkaline phosphatase is that its reaction rate is linear which means that the reaction rate can be improved with the increase in the incubation time. But the problem is that it gives high background leading to low signal-to-noise ratio. On the contrary, HRP gives lower background. Therefore, alkaline phosphate is more sensitive, while HRP is more specific in its action on the blot. Another type of labeled secondary antibodies popular in use is biotinylated antibodies. This includes a two-step biotin-streptavidin system (Fig. 5). The secondary antibodies labeled with biotin are made to bind to the primary antibodies. Subsequently, streptavidin labeled with HRP or dye or fluorophore is added. This results in an irreversible interaction of biotin with streptavidin and gives signal for detection. Further, gold particles can also be attached to secondary antibodies. Using this, the proteins can be stained in dark red color due to accumulation of gold particles.

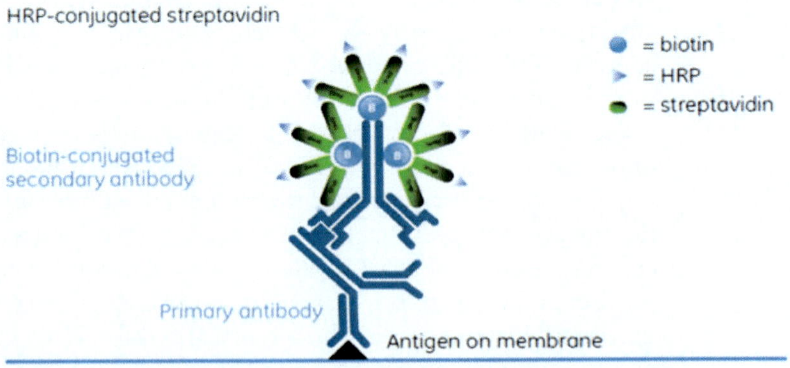

Fig. 5 Biotinylated secondary antibody reaction with streptavidin labeled with HRP [2]

4 Applications in Dairy Science

Some of the important applications of western blotting include:

- Purification of proteins: To find out/confirm the presence of the protein of interest in the sample fractions and to detect impurities before scaling up of a protein production process. The technique can be used to get information regarding milk protein expression in healthy and mastitic animals. Around 2971 milk proteins can be identified using western blotting with false discovery of 0.1% (including proteins from whey, milk fat globular membrane and exosomes from healthy and mastitic animals) [6].
- Detection of low abundance proteins: As this technique is very sensitive, so it can measure a wide range of proteins. The incidences of post-translational modifications like phosphorylated isoforms can be detected. Moreover, fluorescent western blotting can be used for signal amplification for the detection of extremely low abundant proteins. In this, biotinylated secondary antibody is applied followed by streptavidin conjugated with fluorophore.
- Western blotting has found many applications in milk proteomics as well, for example, immunological cross reactivity between milk proteins of different species [7] can be tested using this and such studies have found application in detection of milk protein allergies. Another application is in determining and even quantification of changes between milk proteins from healthy and mastitic animals. Further, detection of adulteration of milk with cheese whey can also be done using western blotting. Glycomacropeptide, a C-terminal peptide released from κ-casein by the action of rennet during cheese processing is the key peptide which comes to the whey and its detection in the milk sample by western blotting [8, 9] confirms the event of adulteration.

References

1. https://info.gbiosciences.com/blog/how-to-prepare-samples-for-western-blot-analysis-1
2. Healthcare GE (2011) Western blotting: principles and methods. In: Handbooks from GE Healthcare. Sweden
3. Kurien BT, Scofield RH (2015) Western blotting: methods and protocols. Springer, Heidelberg
4. https://www.cytivalifesciences.com/en/us/news-center/5-top-western-blot-transfer-problems-and-solutions-10001
5. https://www.researchgate.net/publication/50268856_An_Overview_of_Western_Blotting_for_Determining_Antibody_Specificities_for_Immunohistochemistry/figures?lo=1
6. Reinhardt TA et al (2013) Bovine milk proteome: quantitative changes in normal milk

exosomes, milk fat globule membranes and whey proteomes resulting from Staphylococcus aureus mastitis. J Proteome 82:141–154

7. El-Agamy EI et al (2009) Are camel milk proteins convenient to the nutrition of cow milk allergic children? Small Rumin Res 82(1):1–6
8. Chávez NA et al (2008) Detection of bovine milk adulterated with cheese whey by western blot immunoassay. Food Agric Immunol 19 (4):265–272
9. Neelima S, Rajan Y, Rajput S, Mann B (2013) Chemical and functional properties of glycomacropeptide (GMP) and its role in the detection of cheese whey adulteration in milk: a review. Dairy Sci Technol 93(1):21–43

Chapter 6

Membrane Processes

Abstract

Membrane processing is a different type of technique in which the concentration and separation of the molecules in the solution or sample can be done without application of heat. The separation of the particles is based on their molecular shape and size along with the combination of pressure and semipermeable membranes. This chapter covers the basic aspects of the membrane technology including the design of various types of membranes. Advantages and disadvantages of the techniques have been discussed. Membrane processes like Reverse Osmosis, Nanofiltration, Ultrafiltration, Microfiltration, Dialysis, Gas permeation, Pervaporation, and Electrodialysis have been covered in detail. Emerging concept of liquid membrane along with its applications is also discussed. One of the important applications which is to fractionate milk proteins by ultrafiltration is also covered.

Keywords Reverse osmosis, Nanofiltration, Ultrafiltration, Microfiltration, Dialysis, Gas permeation, Pervaporation, Electrodialysis, Liquid membrane

1 Basic Principle

Membrane processing is a different type of technique in which the concentration and separation of the molecules in the solution or sample can be done without application of heat and other harsh sample processing conditions. The separation of the particles is based on their molecular shape and size along with the combination of pressure and semipermeable membranes (designed for this technique). The word "membrane" originates from a Latin word "membrane" meaning "skin." Thus, a membrane can be described as a selective barrier that divides two phases and selectively restricts the transfer of various compounds. The makeup of the membrane varies as it can be thin/thick, structurally it can be homogeneous/heterogeneous and the mode of transportation or separation of the particles can be active/passive. When the mode of transportation is driven under the influence of temperature difference, concentration, or pressure it is termed as passive transport. The origin of the membrane can be from natural sources, that is, biological in nature or synthetic, it can be charged or neutral in nature. The membrane is otherwise described as a structure whose sides are significantly

Kamal Gandhi et al., *Advanced Analytical Techniques in Dairy Chemistry*, Springer Protocols Handbooks,
https://doi.org/10.1007/978-1-0716-1940-7_6,

larger than its thickness, through which mass transfers may take place under a number of driving forces.

Separation processes of fluid mixtures can be categorized as equilibrium separation and rate governed separation. Equilibrium separation is a process in which the product phases are in equilibrium with the feed phases. A system is said to be at equilibrium, if it is at constant pressure, temperature, and composition. The system is stable, not changing with respect to time. So, at this state the chemical potential of both the phases is equal. So, in the rate governed separation, the separation actually happens because of the difference in the rate of physical transport of the particles and that is the transport of component from higher concentration to lower concentration, using some medium and under the influence of a driving force. Most of the membrane processes actually come under rate governed separation process. Equilibrium governed separation includes distillation, absorption, adsorption, drying, and under the rate governed separation process comes the osmosis, reverse osmosis, dialysis, filtration or microfiltration, etc.

Membrane separation device consists of a module which houses the membrane. The membrane is considered as the heart of a membrane process and is generally a semipermeable barrier or interface between the two phases. The semipermeability of the membrane means that the membrane will allow certain solutes to be retained on its surface, while the remaining part will pass through it as the permeate. Usually, the solutes or ions get retained, while the solvent passes through. The driving force can be either pressure, concentration, temperature, or electromotive forces or a combination of these which will result in the separation of the components.

Feed phase is called as upstream side phase, while permeate phase as downstream phase. In any membrane separation process, separation is achieved, due to the ability of the membrane to transfer one component of the feed at a faster rate than the other components. So, performance of the membrane is given by the two parameters. First one is selectivity and the second one is flow through or flux of the membrane. The volume of the liquid or sample flowing through the membrane per unit area and time is termed as flux. Retention and separation factors are used to express the selectivity of a membrane towards a liquid or mixture. Retention which is denoted by the 'R' actually varies from zero percent to hundred percent. Hundred percent retention means that there is a complete retention of the solute, everything is getting retained on the surface of the membrane. Zero percent means that nothing is retained by the membrane. The selectivity of the membrane for the gaseous mixture and mixtures of organic liquids is expressed as separation factor 'α'. The value of selectivity alpha is chosen in such a way that it is usually greater than unity. Hence, it is denoted by α A/B, that is, when the permeation rate of component A exceeds than that of component B. Similarly, when B exceeds A, then it is written as α B/A. When the value of α A/B is equivalent to α B/A, it signifies that no separation is achieved. For a separation to be achieved, the separation factor, alpha should be always more than 1.

2 Advantages of Membrane Separation Process

1. Energy consumption is very low. Membrane processes do not require any energy except that is required to supply the feed or pump the feed from the feed tank to the membrane module.
2. Membrane processes can be hybridized very nicely with another process.
3. Upscaling is easy for membrane process.
4. The properties of the membrane can be varied or adjusted so that it suits to one's requirement.
5. Membrane processes do not require additives which will save the extra cost.
6. Fractionate the component in their native state.

3 Drawbacks of Membrane Technology

1. Concentration polarization is a phenomenon in which the solutes gets deposited on the surface of the membrane. As the membrane process is continuous, this will increase, more and more and the solutes will get deposited on the surface.
2. External or internal fouling: External fouling occurs when the solute gets deposited on the surface of the membrane and subsequently blocks its pores. In internal fouling, the solutes block the internal passage and will not move out of the column. So, both the external or internal fouling are not good for the membrane processes.
3. Usually the lifetime of membrane is less than other separation processes.
4. Low selectivity or the flux: Flux will be high initially and goes on decreasing with time due to concentration polarization or the fouling.
5. Upscaling is not so good because of linearity.

So, nanofiltration and reverse osmosis actually overlap each other in terms of driving force as well as separation size range. All the membrane processes like ultrafiltration, nanofiltration, microfiltration, reverse osmosis are pressure driven membrane processes

Table 1
Membrane processes with their driving forces and separation size range

Name of process	Separation size range	Driving force
Microfiltration	10–0.1 μm	Pressure gradient
Ultrafiltration	Less than 0.1 μm to 5 nm	Pressure gradient
Reverse osmosis	Less than 5 nm	Pressure gradient
Electrodialysis	Less than 5 nm	Electric field gradient
Dialysis	Less than 5 nm	Concentration gradient

as they are operated under the gradient of pressure. As one proceeds from microfiltration to ultrafiltration, reverse osmosis and electrodialysis, dialysis the size of the solutes that will be separated is usually decreasing. Table 1 shows the membrane processes along with their driving forces and separation size range.

For separating the smaller solutes, we still require higher pressure. So, the pressure increases as one goes from microfiltration to nanofiltration. Both the physicochemical properties of the solutes and the physicochemical properties of the membranes which are involved are essential to achieve a particular separation. The separation of the components of the feed occurs due to the driving force and a direct relation exists between the permeation rate and the driving force.

A normal filter can also be considered as a membrane by definition. A normal filter is generally a usual filter, cloth filter or any such filter that is available at home or labs. However, the filter is generally restricted by convention to structures which can distinguish suspension particles of between 1 and 10 μ. The basic difference between the conventional filter and membrane includes the sizes that is getting separated.

4 Principal Types of Membranes

Isotropic membrane, anisotropic membranes, ceramic or inorganic membranes, and liquid membrane are the principal types of membranes.

4.1 Isotropic Membrane

Under isotropic, we can have microporous membrane, nonporous, dense membranes, and electrically charged membranes. They have a uniform composition throughout with being either porous or nonporous. The total membrane thickness determines the resistance created by the membranes to mass transfer. Increased permeation rates can be achieved by reducing the membrane thickness.

4.1.1 Microporous Membranes

These are much similar to the conventional filter with very small pore sizes. The membrane totally rejects any particles larger than the largest pores, that is, solutes with a size greater than the pore size of the membrane get retained, while the particles whose size range falls between the largest and the smallest pore are partially rejected on the basis of the pore size distribution. Membranes do not have uniform pores, they have a pore size distribution. The type of membranes falling under this category is microfiltration and ultrafiltration.

4.1.2 Nonporous Dense Membrane

Membrane consists of a thick film through which diffusion under heat, concentration, or the electrical potential gradient takes place. These membranes do not have pores and hence referred as nonporous membranes. The separation depends on the relative transport rates of the components or the particles (to be separated) within the membrane. Solutes which are supposed to be transported must be soluble in the membrane and they should have a proper diffusion rate so that they will be transported across the membrane. As a result, if the concentrations of permeates in the membrane content are significantly different, these membranes can differentiate between permeates of similar size. These membranes, unless they are very thin, have the disadvantage of providing a low flux. As a consequence, such membranes are inserted into the top skin layers of the asymmetric membrane. Dense membranes are often used in processes such as reverse osmosis, pervaporation, and gas isolation.

4.1.3 Electrically Charged Membranes

These membranes can be dense or it can be microporous or both but most are commonly very finely microporous. Most of the electrically charged membranes are microporous with the walls getting fixed positive or negatively charged ions. Positively charged membrane or anion exchange membrane have positively charged ions fixed on it, while negatively charged or cation exchange membranes have negatively charged ions. The separation is independent of the pores size of the membrane and takes place on the basis of the concentration and charge possessed by the ions.

4.2 Anisotropic Membranes

The membranes are composed of many layers, each of which has various structures and permeability. A standard anisotropic membrane has a comparatively thin (skin layer) layer, a selective layer which is backed on an open, much thicker, porous substructure. The surface layer determines the permeation rates and separation properties. The below layer which is mostly porous than the surface layer is providing the mechanical support and virtually has nothing to do with the separation. The thin surface layer, that is, the skin layer determines the resistance to mass transfer. The flux is very high that almost all commercial processes nowadays are using this type of membranes.

4.3 Inorganic Membranes

These membranes are also referred to as ceramic membranes. These are very versatile and can be operated at elevated temperature that is the beauty of this ceramic membranes that they can withstand higher temperature, almost close to thousand degree centigrade and they can withstand higher pH, they are high chlorine resistance, their mechanical stability and chemical and thermal stability are very good. These membranes possess asymmetric structure which is composed of at least two or three different porosity levels. So, inorganic membranes offer potential application such as air separation by mixed oxygen ionic and electronic conducting ceramic membranes and molecular sieve.

4.4 Gas Separation/ Permeation

The liquid membrane is a latest addition to the membrane group. Its application is increasing after the facilitated transport came into picture. It uses carriers to transmit components including metal ions selectively through the membrane interface at a comparatively fast rate. It consists of a fluid phase, a thin oil film that exists as a boundary membrane between two phases of solvent or gas mixture in either assisted or unsupported form. These membranes may also be used in a pilot plant for the targeted elimination of organic solvents and heavy metal ions in toxic waste. Isolation and purification of analytes at the laboratory can also be done due to the availability of commercial laboratory scale devices.

5 Membrane Processes

Four basic pressure driven membrane processes, viz. microfiltration, ultrafiltration, nanofiltration and reverse osmosis are depicted in Fig. 1. The microfiltration membrane will allow sugars, proteins, salts, and low molecular weight molecules to pass through the

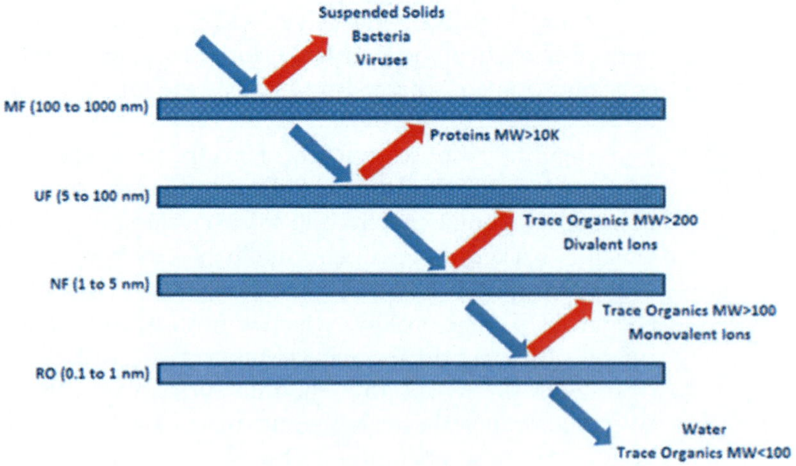

Fig. 1 Pressure driven membrane processes (Source: Lenntech [1])

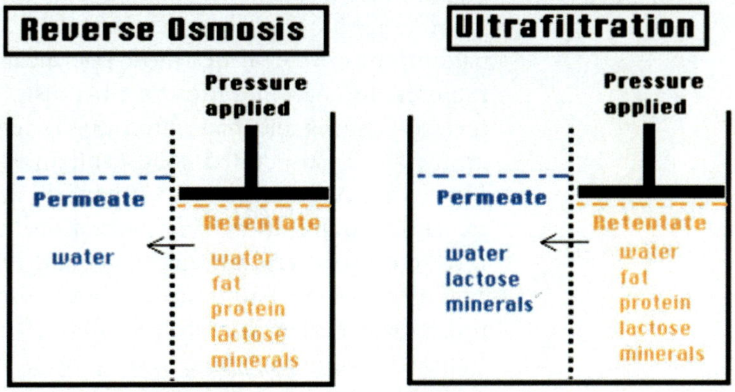

Fig. 2 Process of reverse osmosis and ultrafiltration

permeate side, whereas it will retain bacteria and suspended solids. In ultrafiltration, all these will pass through; however, it will reject viruses, bacteria as well as suspended solids. In nanofiltration, again some of the monovalent ions depending upon the sizes will be retained and some will pass through. It will reject virus, bacteria, and suspended solids. In reverse osmosis only water or solvent can pass through all other ions, viruses, bacteria, any suspended solid will be returned. The pore size of the membranes decreases in the order of: microfiltration > ultrafiltration > nanofiltration and finally the reverse osmosis.

These are categorized as follows, depending on the mode of action and the form of membranes used:

5.1 Reverse Osmosis

Reverse osmosis (Fig. 2) is a membrane separation mechanism that separates the liquid from the other components of a fluid through a membrane under the effect of a pressure gradient. While the liquid flows across the membrane, the solutes are partly or entirely retained. The material of the membrane and the structure of the membrane layer play an important role in the separation of the solute and the solvent permeability. Application of transmembrane pressure to concentrated solutions forces solvent through the RO membrane towards the lower concentration. Important applications of reverse osmosis are sea water desalination, wastewater treatment, and ultrapure water production. All the things like suspended solids, bacteria, viruses, multivalent and monovalent ions will be rejected by the RO membrane and only solvent will flow. The membrane configuration is usually cross-flow. The pore size of the membrane used in reverse osmosis is very small, which allows only minute quantity of very low molecular weight solutes to pass through the membranes. Reverse osmosis is considered as a concentration process, employing the use of 100 MW cut-off membranes with the membrane configuration being of cross-flow type [2].

5.2 Nanofiltration

Nanofiltration is almost similar to that of reverse osmosis; however, membranes are slightly more permeable. Since the membranes of nanofiltration are little more permeable, hence the pressure required for nanofiltration is obviously lesser than that used in reverse osmosis and nanofiltration. It separates at molecular level removing all suspended solids and most of the dissolved solids. Nanofiltration spans the gap in particle size between reverse osmosis and ultrafiltration. Nanofiltration lies somewhere between reverse osmosis and ultrafiltration and that is why it is also called as ultra-osmosis. The transmembrane pressure for this nanofiltration usually ranges from 40 to 200 psi. The applications of nanofiltration include sea water desalination, to concentrate sugar and clarified sugar stream in sugar industry, deacidification of whey during the cheese production and recovery of dyes in textile industries, separation of heavy metals from water and waste water, removal of sulfate phosphate nitrate and fluoride. Membrane will allow only water and salts to pass through, whereas all other dissolved organics bacteria, viruses any other macromolecular organics and particles and silt will be retained or rejected by it.

5.3 Ultrafiltration

Ultrafiltration (Fig. 2) is used to separate large molecules of solutes, the molecule size is usually larger than that of the microfiltration and separates large molecule from smaller molecule and large molecules from other molecules. The primary mechanism is size exclusion. But chemical interactions between solute and membrane and other operating parameters can also affect the process. Two important things which are to be considered during the ultrafiltration process are the physicochemical properties of the solute and the membrane. Normal transmembrane pressure for the ultrafiltration ranges from 10 to 100 psi. Ultrafiltration is a membrane separation process, in which fractionation of a solution or liquid occurs as a function of their solvated size and structure under the influence of the pressure gradient. Usually configuration of the membrane is of cross-flow type. Ultrafiltration membranes have a greater pore capacity, which enables some materials to flow through the pores along with the water. It is a separation/fractionation method that makes use of 10,000 MW cut-off membranes. Ultrafiltration of milk enables a 50% isolation rate of lactose and minerals: the composition of the retentates, for instance, will be 100% fats and the proteins with 50% lactose and 50% free mineral. Diafiltration, where retentate is mixed with water and filtered back, is another form of ultrafiltration procedure, reducing the concentration of soluble permeate compounds with a concomitant increase and a greater amount of the retained materials [3].

5.4 Microfiltration

Microfiltration can be defined as a pressure driven process which is used to separate micron sized particles from fluid. There are two types of microfiltration configuration, namely one is cross-flow and dead-end filtration. The former is advantageous for higher concentrations since deposits on the membrane are washed away by parallel flow through the membrane surface, while the latter is advantageous for solutions with a relatively low solid material. It is only used at a laboratory scale to characterize certain membranes and to develop a process. Microfiltration is identical to ultrafiltration, but the membrane pores are wider, allowing particles with a diameter of 0.2–2 μm to migrate under the effect of a lower pressure than that used in ultrafiltration. Transmembrane pressure which used in microfiltration is usually 1 to 50 psi. Membranes in cross-flow configuration are generally available. This technique enables the preparation of cell-free extracts from a variety of fermented milk products for subsequent biochemical study.

5.5 Dialysis

It is usually used for the concentration purposes, in which the dissolved solute(s) flow across the membrane due to the differences in the concentration in inside and outside of the membrane. Transfer of the material across the membrane occurs through diffusion in such a way that smaller molecules diffuse at a rapid rate than the larger ones. Smaller molecules diffusivity will be higher than that the larger molecules, so the rate of diffusion plays an important role in separating these molecules. From the dialysis membrane, minerals like sodium, potassium, calcium can pass through creatinine and urea which are large molecular components that will be rejected by the dialysis membrane (Fig. 3). The application of dialysis includes removal of salt(s) and other low molecular weight compounds from solution of macromolecules, removing the alkali(s) and acid(s) from a product, separation of alcohol from beer, concentration and hemodialysis.

5.6 Gas Permeation

Gas permeation or gas separation is one of the most important membrane separations and has lot many industrial applications. Porous or dense membranes are used for separation of gas mixtures. Generally, transportation through the membrane occurs due to the phenomenon called as Knudsen diffusion, solution diffusion, or molecular sieving (Fig. 4). When the separation is based on the

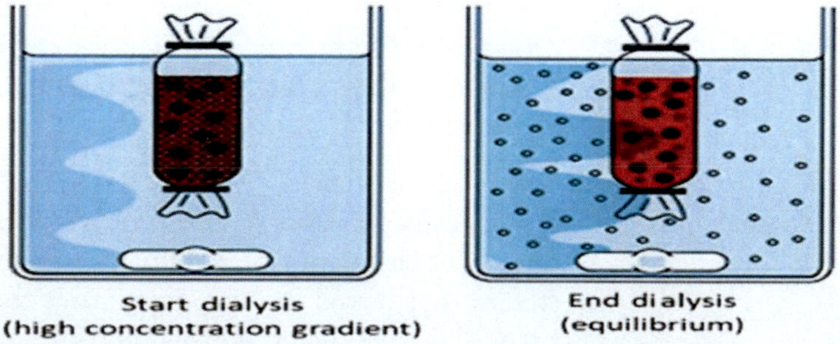

Fig. 3 Dialysis process [4]

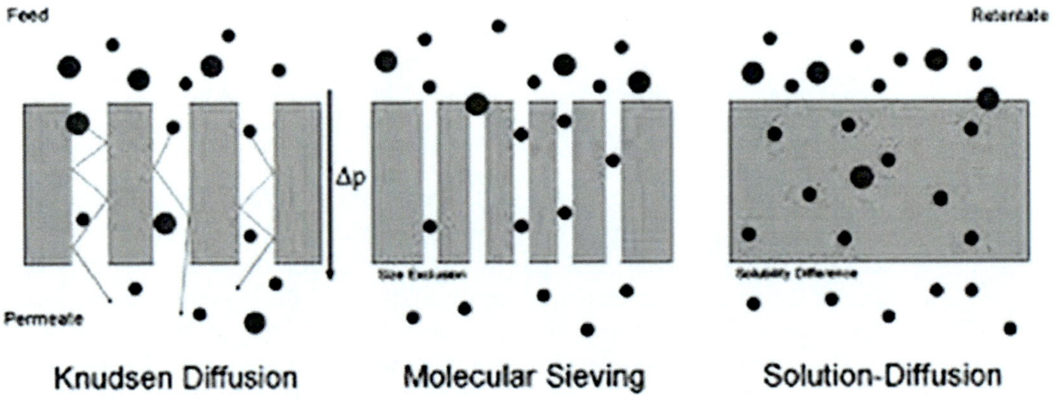

Fig. 4 Gel permeation through membranes [5]

inverse square root of ratio of the molecular weight of separating components A and B, it is called as Knudsen flow. The next one is molecular sieving in which the separation is based on diffusion and sorption characteristics. So, the rate of diffusivity of the solutes plays a substantial role in this type of process. In case of separation by solution diffusion, it is on the diffusivity and solubility of the solute.

Gas separation has lot many applications: Hydrogen recovery from synthesis gas, oxygen enriched air for combustion and medical applications, nitrogen enrichment for wafer-fabrication, metallurgy and inerting, carbon dioxide for enhanced oil recovery, and finally helium for natural gas.

Advantages and disadvantages of using membranes in gas absorption are:

- It provides large surface area per unit volume which is very important because to have a better mass transfer we should have more surface area. In a smaller place we can have more surface area by providing this type of membranes as against conventional absorption system.
- No flow constraints such as flooding or loading existed as those exist in usual conventional absorption processes.
- One of the disadvantages is that the membrane thickness will provide a resistance to mass transfer; however, if we can have a very thinner membrane, then we can almost nullify their resistance to flow which is being provided by the thickness of the membrane.

5.7 Pervaporation

It is a permeate vaporization process wherein a liquid mixture is in direct contact with one side of the membrane and the permeate stream is removed in vapor state from the other side of the membrane (Fig. 5). The permeate is getting vaporized, so pervaporation

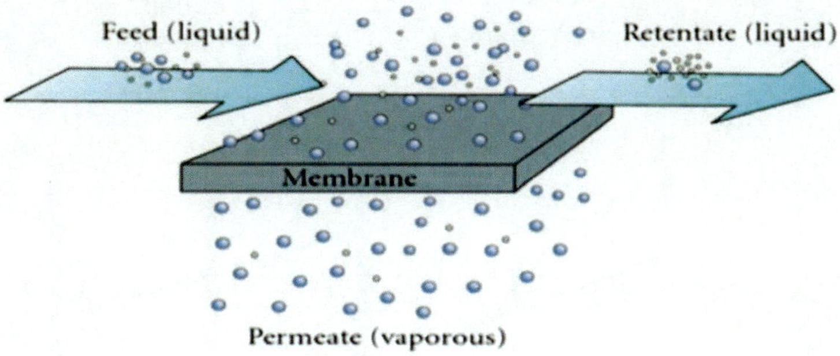

Fig. 5 Principles of pervaporation [6]

separates usually a low volatile or low boiling point liquid from a nonvolatile liquid. Solution diffusion is the mechanism of the separation. Pervaporation is sometimes used to split or separate azeotropic mixtures, to separate heat-sensitive materials, and to dehydrate and then remove volatile organic compounds from polluted water.

5.8 Electrodialysis

It is a dialysis process which is happening under an electric gradient, so electrodialysis is an electrochemical process used to separate charged particles from aqueous solutions or from neutral solutes. It is the movement of ions through ion-selective membranes under the influence of an electromotive force applied across the membrane area. Cations with a positive charge migrate to the cathode, whereas anions with a negative charge transfer to the anode (Fig. 6). Applications of the electrodialysis include desalination of water, purification of organic compounds, wastewater treatment, wine stabilization, where wherein the tartaric acid is removed during the fermentation and also in recycling and regeneration of galvanic baths in the metallurgical industries.

5.9 Liquid Membranes

Hydrophobic solvents function as selective barriers between the aqueous feed solution and the absorption phase in these types of membrane processes. BLM, SLM, and ELM are three different types of the liquid membranes (Fig. 7). BLM stands for the bulk liquid membrane, SLM stands for the supported liquid membrane, and ELM stands for the emulsion liquid membrane. Bulk liquid membranes are not commercialized, they are utilized in the lab scale to generate data and to see whether the process of liquid membrane is feasible for a particular separation or not. SLMs and ELMs are already adapted in various industrial sectors. SLM is a supported liquid membrane in which a liquid (membrane) is impregnated inside the pores. Separation will be achieved in the pores only; the membrane itself does not do anything but it just provides the mechanical support or it just holds the liquid membrane phase. In case of the ELMs, the solute will go to the receiving

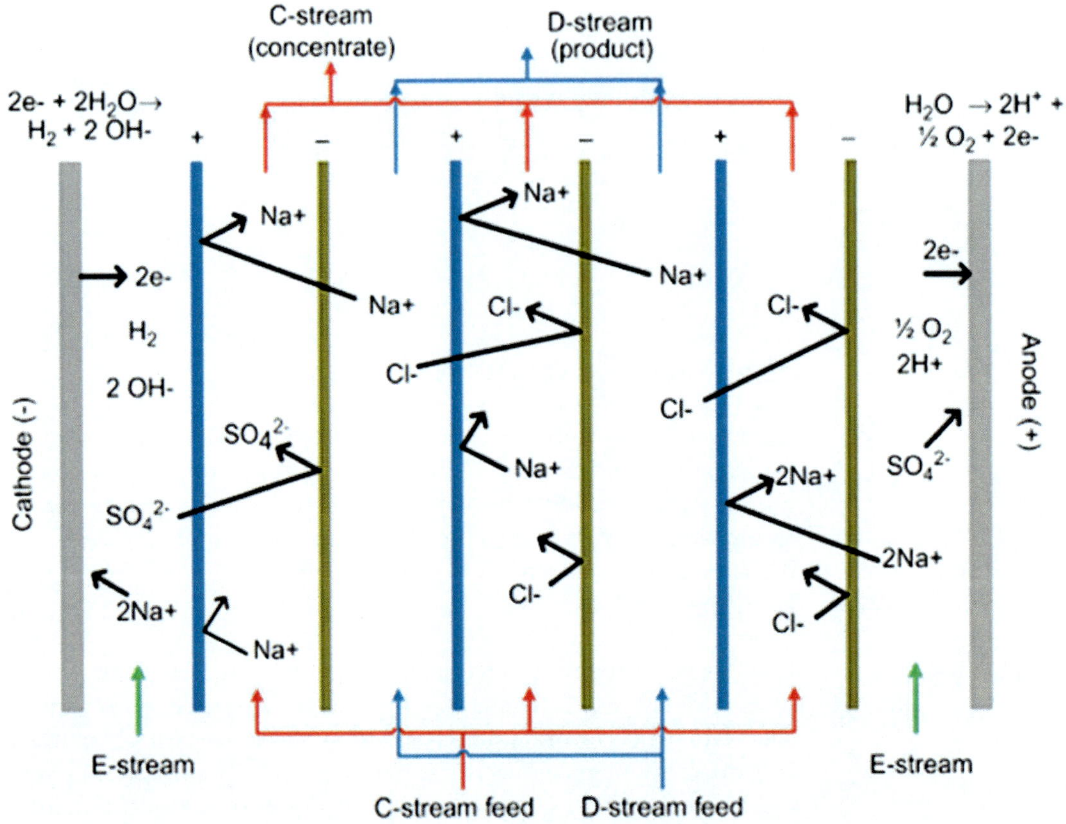

Fig. 6 Electrodialysis [7]

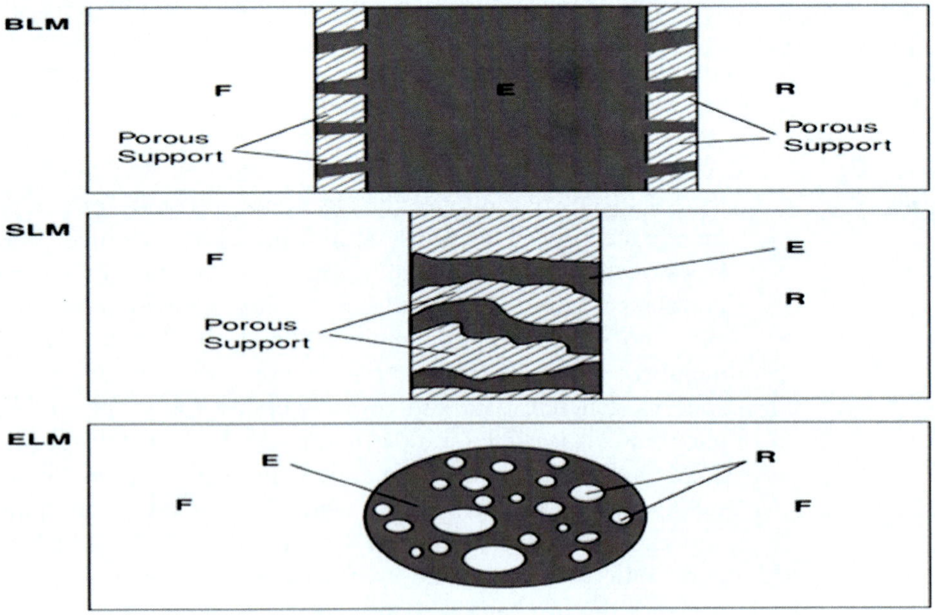

Fig. 7 Configurations of the liquid membrane systems [8]

phase from the feed phase. They will then encounter emulsifier breakers which will break the emulsifier and the desired solute will get separated from the feed stream. The major advantage actually of the ELM is that we can create very high membranes specific surface area, almost 1000–3000 m^2/m^3. The application of liquid membrane includes removal of heavy metals from effluents, dephenolation of wastewater that means removing phenol from various water system, separation and concentration of amino acids, enzymatic bioconversions, and gas separation.

6 Fractionation of Milk Proteins by Ultrafiltration

6.1 Principle

The separation of particles/macromolecules/molecules by ultrafiltration is based on the differences in their size where in the solution or the sample is driven through a semipermeable membrane under the influence of a transmembrane force like pressure or centrifugation. The pores of the membrane range in size from 1 to 20 nm, and the pore size of the membrane is chosen in such a way that the protein/peptide of interest is too large to pass through. Membranes with molecular cut-off values of 500–1,000,000 Da and an operating pressure range of 10–200 psi are typically used. Numerous apparatuses are available for use on a small scale that utilizes centrifugal force to separate molecules passing through a semipermeable membrane. For handling small volume ranging from 1 to 400 ml at the laboratory level, stirred cells are the best. Polycarbonate, regenerated cellulose, polyvinyl difluoride, polyvinyl chloride, polyacrylonitrile, polyamide, polysulfone, and cellulose acetate are the most often used ultrafiltration membrane materials. Certain inorganic-based membranes incorporate a carbon support sheet, while the ultrafiltration layer is made of zirconium oxide. These membranes have an incredibly high thermal and chemical tolerance, can be worked at temperatures up to 100 °C, and can be sterilized in an autoclave.

Nowadays, advances in the properties of membranes such as their configuration, composition, and porosity have enabled the discovery of novel methods for fractionating milk proteins into their constituents. Separation of all milk proteins is theoretically possible by selective membrane filtration of skim milk. The milk proteins are separated in the following order: lipoproteins in MFGM > casein micelles > immunoglobulins > lactoferrin > bovine serum albumin > beta-lactoglobulin > alpha lactalbumin > milk protein dependent peptides (separation being on the basis of size largest to the smallest). Ultrafiltration has been shown to be a more effective method of fractionating and concentrating milk proteins and their related peptides. Ultrafiltration may be used to fractionate whey proteins and bioactive peptides for medicinal, biomedical, and dietary purposes.

6.2 Materials

1. Membranes of different cut-off limits.
2. Nitrogen gas cylinder.
3. Stirred cell.
4. Magnetic stirrer.
5. Centrifuge.
6. Centrifugal devices fitted with different cut-off membranes.

6.3 Procedure

6.3.1 Centrifugal Ultrafiltration

1. Position sample reservoirs into filtrate vials.
2. Add the sample reservoir's solution (2 ml maximum volume). Avoid touching the membrane top with the pipette cap. Utilize a covered rotor wherever possible to mitigate sample evaporation. Alternatively, you can secure the instrument by connecting the retentate vial to the sample reservoir. The rotor adapter must accommodate a rotor of at least 137 mm in duration.
3. Insert the covering gadget with the connected filtrate vial into the centrifuge rotor; balance with another identical device. Centrifuge the contents at 4000–7500 $\times$ *g* if the retentate vial is in place or in absence of the retentate vial centrifuge at 4000–6500 $\times$ *g* to the desired concentration.
4. The centrifugal filter assembly is removed from the centrifuge. Preserve the filtrate for filtration, while the filtrate can either be discarded or left in the filtrate vial for concentration.
5. Recover the retentate by placing the retentate vial on the sample reservoir.
6. In order to transfer the concentrate into the retentate vial, centrifuge the contents at 300–1000 $\times$ *g* for 2 min.
7. The retentate vial should be removed from the centrifuge and separate it from the concentrator. Cover the retentate or the filtrate vial using cap cover.

6.3.2 Ultrafiltration Using Stirred Cells

1. Pretreatment of the ultrafiltration membrane for removing the preservatives should be done as per the manufacturer's instructions. Generally the membrane should be soaked in water thrice, that is, the water should be replaced every time.
2. Follow the manufacturer's instructions to assemble the stirred cell. The membrane should be placed in correct manner, that is, shiny surface uppermost.
3. In order to remove the particulate matter, the solution should be pre-filtered or centrifuged. Pour this solution in the stirred cell and place the assembly on the retaining stand.
4. The pressure relief cap should be closed and a nitrogen cylinder which acts as a regulated pressure source should be attached to it. The entire assembly should be placed on the magnetic stirrer.

5. Apply the least amount of pressure necessary to maintain the desired flow rate; higher pressures will result in increased fouling and concentration polarization, resulting in a decreased flux rate. The applied pressure does not surpass the manufacturer's maximum pressure specification. Collect the filtrate in a different vessel.
6. Change the magnetic stirrer's stirring intensity such that the vortex is just one-third the depth of the solution. Excessive stirring results in shearing forces and foaming, which results in protein denaturation, while insufficient stirring results in enhanced concentrate polarization.
7. As the desired level of concentration is achieved, the pressure should be turned off by opening the pressure release valve. The stirring should be continued for 5 min for resuspending the polarized layer.
8. The cap assembly should be removed and the concentrate should be poured gently.
9. The above measures may be replicated for separate membranes regardless of their fractionation cut-off limits for milk proteins/peptides.

References

1. https://www.lenntech.es/processes/mbr-introduction.htm
2. Mann B, Sangwan R, Kumar R (2008) Research techniques in dairy chemistry. NDRI, Karnal
3. Sharma R (2011) Membrane filtration. ozscientific.com 8(8):1782–1789
4. Onnainty R, Granero G (2019) Chapter 14—chitosan-based nanocomposites: promising materials for drug delivery applications. In: Grumezescu AM (ed) Biomedical applications of nanoparticles. William Andrew Publishing, pp 375–407
5. Ismail AF, Khulbe KC, Matsuura T (2015) Fundamentals of gas permeation through membranes. In: Ismail AF, Chandra Khulbe K, Matsuura T (eds) Gas separation membranes: polymeric and inorganic. Springer, Cham, pp 11–35
6. Mixa A, Staudt C (2008) Membrane-based separation of phenol/water mixtures using ionically and covalently cross-linked ethylene-methacrylic acid copolymers. Int J Chem Eng 2008. (1687-806X)
7. Hutten IM (2007) Handbook of nonwoven filter media. Elsevier, Oxford
8. Kislik VS (2009) Liquid membranes: principles and applications in chemical separations and wastewater treatment. Elsevier

Chapter 7

Potentiometry

Abstract

Potentiometric analysis is based on the notion of measuring an electrochemical cell's potential when a small amount of current is allowed to flow through it. It is considered to be one of the most accurate quantitative techniques. Besides being accurate, it offers an added advantage of analyzing the colored or turbid solutions successfully. Other advantages of this technique, such as speed and time required for analysis, no or minimal training required to operate the instrument, and lower equipment cost, make it an attractive analytical technique. In case of dairy products, it is useful and employed in determining the pH and acidity, determining the presence of neutralizers, estimation of metals like sodium, potassium, fluoride, calcium, ammonium, and nitrate using ion-selective electrodes (ISEs), and also for the heavy metals detection. In this chapter, the basic principle of potentiometry and various types of electrodes like metallic electrodes, membrane electrodes including glass membrane electrodes, solid state ISEs, liquid membrane ISEs, gas-sensing electrodes, and enzyme electrodes are covered. The principle and working of pH meter and buffer solutions are also included in the chapter. The standard protocol for operating a pH meter is also covered. Application of ion-selective electrodes for measuring the ionic calcium in skim milk has been covered.

Keywords Metallic electrodes, Membrane electrodes including glass membrane electrodes, Solid state ISEs, Liquid membrane ISEs, Gas-sensing electrodes, Enzyme electrodes

Potentiometry is considered to be one of the most accurate quantitative techniques. Besides being accurate, it offers an added advantage of analyzing the colored or turbid solutions successfully. Other advantages of this technique such as speed and time required for analysis, no or minimal training required to operate the instrument, and lower equipment cost make it an attractive analytical technique. In case of dairy products, it is useful and employed in determining the pH and acidity, determining the presence of neutralizers, estimation of metals like sodium, potassium, fluoride, calcium, ammonium, and nitrate using the ion-selective electrodes (ISEs) and also for the heavy metals detection.

Kamal Gandhi et al., *Advanced Analytical Techniques in Dairy Chemistry*, Springer Protocols Handbooks,
https://doi.org/10.1007/978-1-0716-1940-7_7,

1 Principle

Potentiometric analysis is based on the notion of measuring an electrochemical cell's potential when a small amount of current is allowed to flow through it. The concentration of the target species in the electrochemical cell is determined from the potential obtained using the Nernst equation. In other words, potentiometry is the difference in potential between an indicator electrode (E_{cell}) and a reference electrode that depends on the membrane potential, E_m. The observed E_{cell} is mainly contributed by

$$E_{cell} = (E_{m+}E_{refi}) - (E_{refr} + E_J) \tag{1}$$

where Ej, E_{refr}, and E_{refi} denote the potentials of the liquid junction, reference electrode, and indicator electrode, respectively. Another possible cause of E_{cell} is the E_{asym}, which occurs as a result of the difference in potential between the exterior and interior surfaces of the indicator membrane, which fluctuates over time, necessitating periodic electrode calibration. It is calculated and assumed constant and its size is small in relation to the entire E_{cell}.

As a result, Eq. 1 may be rewritten as follows:

$$E_{cell} = K + E_m \tag{2}$$

where K is a constant.

as well as

$$E_m = E_{pb1} + E_{pb2} + E_d \tag{3}$$

where E_{pb1} denotes the phase boundary potential at the membrane's external surface, E_{pb2} is the potential inside the membrane of the indicator electrode, and E_d is the potential created as a result of charge separation. Due to the fact that its value is negligible, it is treated as zero. E_{pb2} may be kept constant by measuring and recording the concentration of ions in contact with the inner membrane. Thus, E_{cell} may also refer to a membrane's ability to be permselective (allowing only single-charged ions to partition into the membrane), resulting in measurable charge separation (E_{pb1}). Assuming each ion has a linear concentration gradient across the membrane and the membrane is permeable to just one ion, we may construct a newer equation known as the Nernst equation:

$$E_{cell} = K + RT\frac{\ln(a_i)}{q\,F} \tag{4}$$

where K denotes a constant, F denotes Faraday's constant, R denotes the gas constant, and T is the temperature in degrees Kelvin. q denotes the charge (including the sign) on the analyte ion and a_i denotes the analyte ion's activity. The potentiometric analysis is based on the Nernst equation. E_{cell} vs log a_i is a linear function covering the electrode's operating range at a temperature of 25 °C and a charge of 1.

2 Potentiometric Electrodes

Potentiometric electrodes are classified as follows:

2.1 Metallic Electrodes

The type of redox reaction occurring at the electrode–solution interface determines the potential of the metallic electrode. For potentiometric analysis, metallic electrodes are further subdivided into three classes:

1. The first type of this electrode is in which the metal is in contact with a solution containing its ion. The concentration of the cationic form of the metal in solution is important for the electrode to respond, so, in case of silver/silver chloride electrode, the concentration of Ag^+ in solution assists in the determination of the potential of the silver wire.
2. When the potential of a first-type electrode reacts to the potential of another ion in equilibrium with Mn^+, it is referred to be a second-type electrode. These are further subdivided into the calomel and Ag/AgCl reference electrodes.
3. The first two types of electrodes develop a potential due to the redox reaction as a result the oxidation state of the metallic electrode changes. These electrodes serve as a source or sink of the electrons in the potentiometric reactions and are termed as redox electrodes. Their use in direct potentiometry is limited because these electrodes generate response against the concentration of more than one ion.

2.2 Membrane Electrodes

Placing a thin-walled glass membrane in the solutions of varying pH leads to the generation of potential which is now termed as membrane potential. This discovery opened the gates for new type of indicator electrodes which are termed as ion-selective electrodes and now they have been developed for numerous ions. Constant developments are being carried out for developing new membrane electrodes.

2.2.1 Glass Membrane Electrodes

Figure 1 shows a typical glass electrode. These electrodes are formed by doping SiO_2 with various chemicals. The most common of such electrodes are H^+ sensitive electrodes or pH electrodes. The H^+ from solution gets exchanged with Li^+ in the glass.

$$Li^+Gl^- + H^+ \rightleftharpoons H^+Gl^- + Li^+$$

Different glass compositions can be made to measure Na^+, Ag^+, K^+, NH_4^+. The first type of the glass electrodes commercialized was

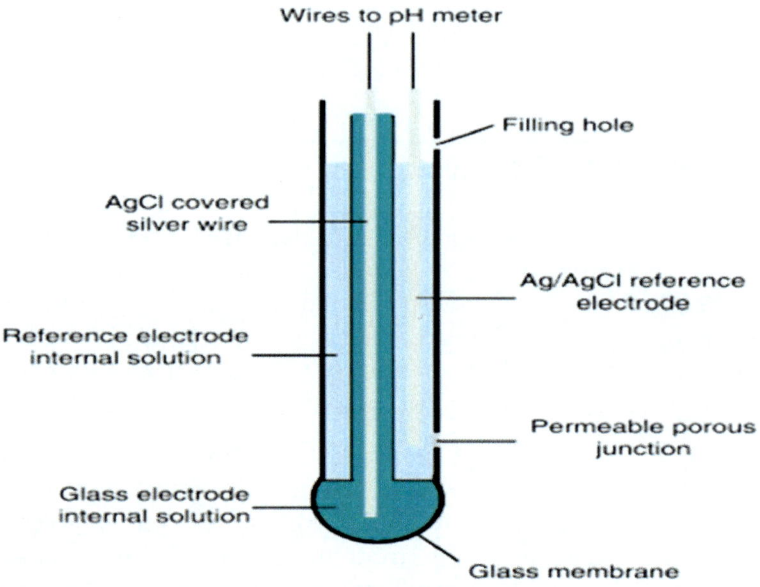

Fig. 1 Glass membrane electrode [1]

manufactured using glass having an approximate composition of 72% SiO_2, 6% CaO, and 22% Na_2O. The outer portion of the membrane, that is, approx 10 nm gets hydrated when the electrode is immersed in an aqueous solution, resulting in the formation of negatively charged sites on the glass, that is, G^-. The counter ions of these negatively charged ions are the sodium ions, which move across the hydrated layer. Another type of ions diffuses from the solution into the membrane, that is, hydrogen ions and they replace the sodium ions as they have a higher binding affinity towards glass than the sodium ions giving rise to the membrane's selectivity for H^+ as shown below.

$$H^+(aq) + G^-\text{----}Na^+(s) \rightarrow G^-\text{----}H^+(s) + Na^+(aq)$$

Charge transfer across the membrane is facilitated by the Na^+ ions. So, the potential of the electrode is generally determined by the difference in the composition of the solution on either side of the membrane. The typical example of glass membrane electrode is the pH electrode. The sensitivity of these electrodes towards other ions can be manipulated by altering the composition of the glass, like an electrode membrane made from 18% Al_2O_3, 11% Na_2O, and 71% SiO_2 is sensitive to potassium.

2.2.2 Solid State ISEs

Solid state ionic conducting membranes are used in these electrodes. These membranes are made from ionic compounds or mixture of ionic compounds. They exhibit a high degree of selectivity since only ions that can penetrate the crystal structure can interfere with the electrode response. This is the fundamental distinction between

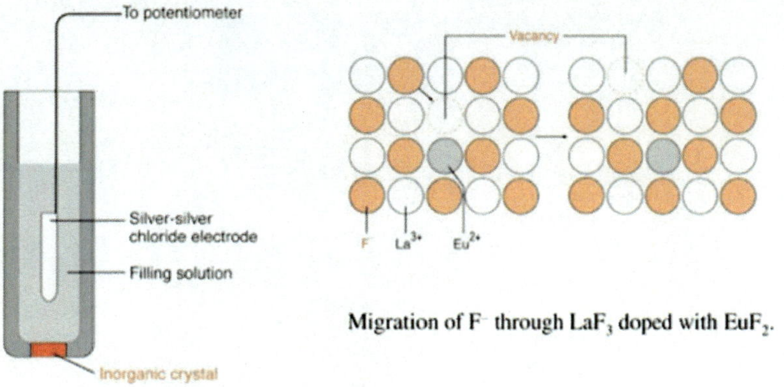

Fig. 2 Solid state crystalline membrane [2]

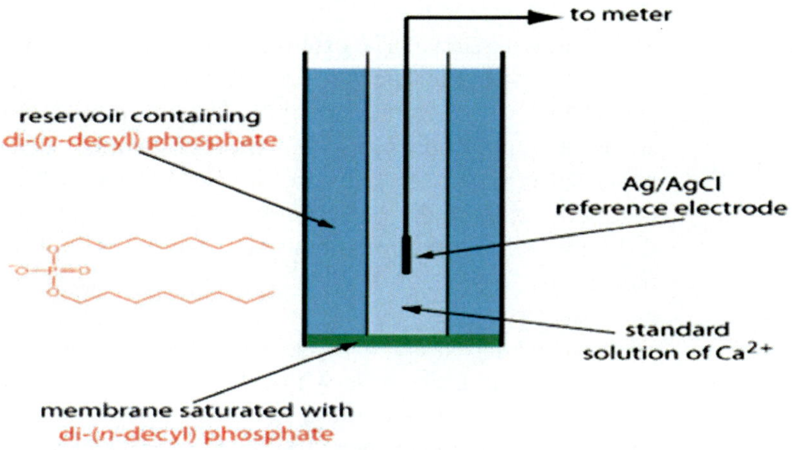

Fig. 3 Liquid membrane ion-selective membrane [3]

these electrodes and those made of glass membranes. Due to the absence of an internal solution, the number of viable junctions are limited. The selectivity of the crystalline membrane may be for both the cation and anion of the membrane-forming components. These electrodes' active membranes are composed of a single inorganic crystal that has been treated with rare earth and integrated into a hydrophobic membrane. The Fluoride selective electrode is an example of this kind of electrode, in which a single crystal of LaF_3 serves as the ion-selective membrane (Fig. 2).

2.2.3 Liquid Membrane ISEs

The membranes of liquid membrane ISE (Fig. 3) contain three major components: an ionophore, a lipophilic solvent, and solid support. In order to provide the solid support, polyvinyl chloride is commonly used, while the ionophore may be an anionic or cationic exchanger or neutral carriers, which are ion specific. The working of ionophores varies from one another, but they generate a potential across the membrane due to the separation of charges resulting in

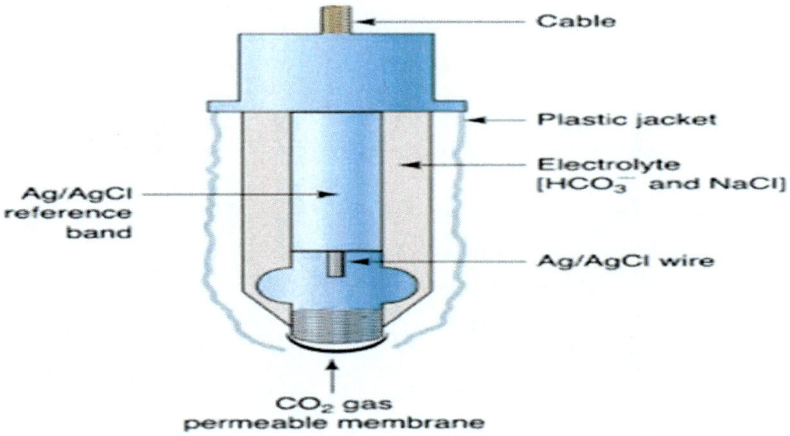

Fig. 4 Gas-sensing electrode [3]

generation of a signal. The extent to which an ionophore complexes selectively with the analyte ion determines the selectivity of an electrode. Nowadays the use of selective lipophilic complexing agents as ionophores has led to the development of a variety of novel selective sensors. In case of Ca^{2+} ion-selective electrode, immobilization of di-(*n*-decyl) phosphate (a chelating ligand) on the polyvinyl chloride membrane is a typical example of a liquid-based ISEs (Fig. 3).

2.2.4 Gas-Sensing Electrodes

Such electrodes consist of a gas permeable membrane (Fig. 4), a combination pH electrode with internal buffer solution. The gas dissolves in the buffer solution which is in contact with the combination pH electrode which results in change in the pH of the solution which is detected by the combination electrode. CO_2, SO_2, NH_3 are measured using these electrodes. Replacing the pH electrode with an ISE enhances the selectivity of the electrode.

2.2.5 Enzyme Electrodes

The principle of an enzyme combining with a specific compound is used for the enzyme electrodes (*see* Fig. 5), and a product that results from this is identified by the appropriate ion-selective electrode. Enzyme is immobilized at the surface of the electrode. Enzymes exhibit a very high degree of selectivity and sensitivity. An enzyme electrode can be made by attaching a thin layer of an enzyme or a biocatalytic material on the surface of an ISE or gas-sensing electrode. Various methods such as covalent attachment on the polymer support of ISE, immobilization on a gel, or direct adsorption on the surface of the electrode are used for attaching the biocatalytic material.

Application of Ions Selective Electrode

- Useful in biology/medicine since it may be used for color or turbid solutions.

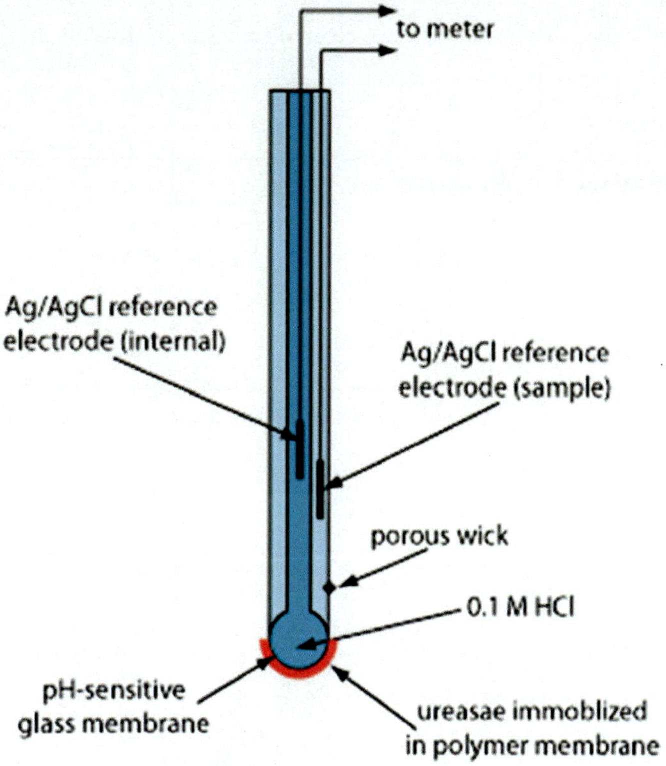

Fig. 5 Enzyme-based electrode [3]

- The change in ion activity over time can be monitored.
- Both the positive and negatively charged ions can be determined.
- They are used in the cosmetic and pharmaceutical industries.
- They are used for monitoring soil water and air.
- They are used in the determination of ions activity and sensors and titrations.

Advantages of Ions Selective Electrode

- Direct measurement of a wide range of ions.
- More precise and generate a good response.
- Efficient and short response time.
- They are relatively cheap and easy to operate.
- They are resistant to solvent/chemical attack.
- They have a wide range of concentration measurement.

Disadvantages of Ions Selective Electrode

- They are unreliable at low concentrations.
- Inability to measure below 2 to 3 parts per million solution.
- Interference by other ions present in the solution.

- The electrode is fragile and has a limited shelf life.
- The electrode can be fouled by protein and other organic solutes.

3 pH Meter and Measurement of pH

A pH system (Fig. 6) has four major parts which are always required.

1. Voltmeter or amplifier to measure small EMF differences.
2. Indicator electrode (pH sensitive): Nowadays the most common type of indicator electrode used in pH measurements is

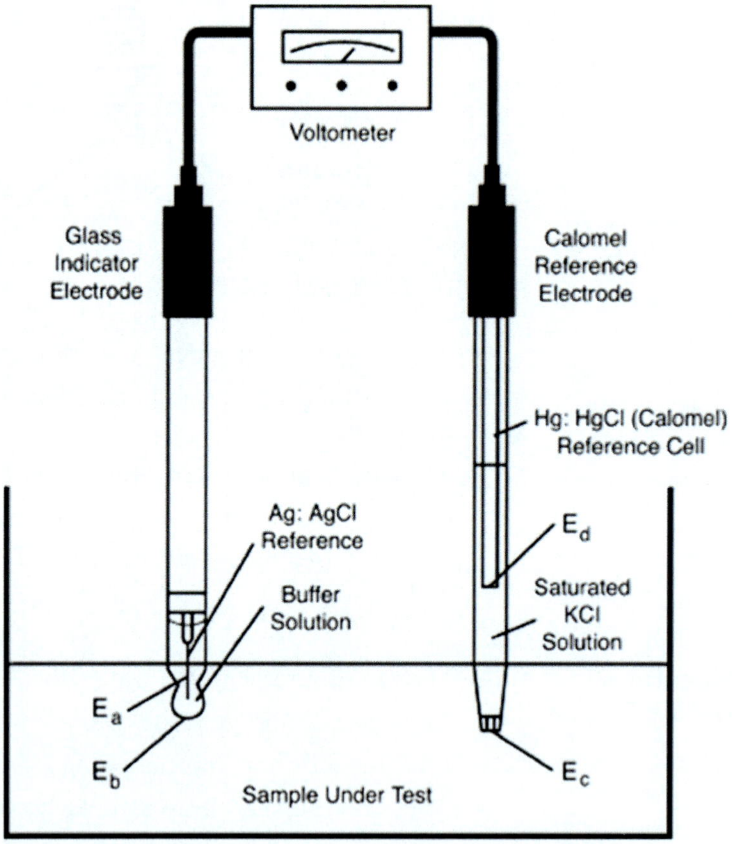

Fig. 6 The potentiometric system along with the measuring circuit. E_a: the potential maintained between Ag/AgCl electrode and the inner liquid. E_b: the potential developed at the pH-sensitive glass membrane. E_c: diffusion potential developed between the test sample and the saturated KCl solution. E_d: contact potential between KCl salt bridge and calomel portion of electrode (Adapted from Nielsen [4])

the glass electrode. An indicator electrode also has three main parts:

(a) Ag-AgCl electrode having a Hg connection which is required as a lead to the potentiometer.

(b) A buffer solution to maintain a constant pH.

(c) A small pH-sensitive glass membrane whose potential varies with the pH of the sample. Measurement of pH using an indicator electrode is carried out against the calomel electrode. These glass electrodes can be operated over a wide range of pH ranging from 1 to 9 but at higher pH values it becomes sensitive to sodium ions. Currently with various advancements, glass electrodes capable of working over the entire pH range are available along with a very minor error due to sodium ions such as <0.01 pH at 25 °C.

3. Reference electrode: The circuit in the pH system is completed with the help of this electrode. This half-cell is one of the most troublesome part of a pH meter. Generally, the problem faced during pH measurement is due to the faulty reference electrode. The most common type of this electrode is the saturated calomel electrode. The reaction occurring in this electrode is reversible in nature:

$$Hg_2Cl_2 + 2e^- \leftrightarrow 2Hg + 2Cl$$

As an indicator electrode, this electrode also has three major parts: (1) platinum wire, covered with a mixture of calomel (Hg_2Cl_2), (2) filling salt solution, that is, saturated KCl solution, (3) a permeable junction through which the filling solution migrates slowly into the sample under measurement. Fibrous or ceramic material is used for making these junctions. The problem with these junctions is that they tend to clog with time leading to unstable response and inaccurate results. The silver–silver chloride electrode is another kind of reference electrode; however, it is less frequent. This electrode performs effectively in situations when the calomel electrode cannot be utilized, such as in samples with a pH greater than 9 or at higher temperatures (80 °C).

This electrode's response is as follows:

$$AgCl(s)^+e^- \leftrightarrow Ag(s) + Cl^-$$

The electrode is made out of a silver-coated platinum wire, with the silver on its surface being transformed to its chloride salt through HCl hydrolysis. To prevent the internal element from being dissolved, the filling solution is a combination of 4 M KCl that has been saturated with AgCl. Porous ceramic is used to make the permeable junction. Because of the increased relative insolubility of AgCl, this kind of electrode is more prone to clogging. This sort of electrode is also available in a twin junction design, in which a secluded inner body retains the Ag/AgCl internal element electrolyte and the ceramic junction, while the outer body, which contains another electrolyte and junction, separates the inner body of the electrode from the sample.

4. Sample to be analyzed.

   pH is usually defined as

$$pH = -\log\ [H^{+}] \tag{5}$$

In other words, it is the response of an electrode to the H^+ ion and, or it is a measure of the H^+ ion activity in a sample.

$$pH = -\log\ [aH^{+}] \tag{6}$$

Using Eq. 5, the pH of 0.1 M HCl will be 1 but the actual pH using Eq. 6 is 1.1. This difference is because the activity coefficient for hydronium ion is not unity in a matrix of 0.1 M HCl. This signifies that the composition of the matrix affects the true pH of a solution. Another complication which is faced during measurement of pH is due to the uncertainty in the relationship between the activity and the potential. The cell potential, E_x, for a solution of unknown pH using a glass membrane electrode is given as

$$E_x = K - \frac{RT}{F} In \frac{1}{a_{H^+}} = K - \frac{2.303RT}{F} pH_x \tag{7}$$

where K includes the reference electrode's potential, the glass membrane's asymmetry potential, and any liquid junction potentials in the electrochemical cell. Because the components that contribute to K change across electrodes and even across days. To correct for this variance, calibration of the pH electrode is required and should be performed using a known-pH standard buffer. The cell potential for the standard, E_S, is

$$E_S = K - \frac{2.303RT}{F} pH_S \tag{8}$$

where pH_S is the pH of the standard. Subtracting Eq. 8 from Eq. 7 and solving for pH gives

$$pH_x = pH_s - \frac{(E_x - E_s)F}{2.303RT} \tag{9}$$

International Union of Pure and Applied Chemistry has approved this as the operational definition of pH.

4 Buffer Solutions

Those solutions that can withstand a small change in pH caused by the addition of a small amount of acid or base. They act by resisting changes in the hydronium and hydroxide ion concentration. Buffer solutions are often composed of a weak acid and its conjugate base, or a weak base and its conjugate acid, with the former being the more favored and prevalent combination. A buffer solution is resistant to pH changes caused by the equilibrium between a weak acid (HA) and its conjugate base (A):

$$\mathrm{HA\ (aq) + H_2O\ (l) \leftrightarrow H_3O^+(aq) + A^-(aq)}$$

4.1 pH of a Buffer

The dissociation constant of the acid in the above equation is given by

$$K_a = [\mathrm{H}^+][\mathrm{A}^-]/[\mathrm{HA}] \tag{10}$$

Taking logarithm on both sides of Eq. 10 gives *Henderson–Hasselbalch equation*, which is the relation between pH and the pKa as follows:

$$\mathrm{pH} = \mathrm{pK_a} + \log_{10}\frac{[\mathrm{A}^-]}{[\mathrm{HA}]} \tag{11}$$

where [HA] and [A^-] are the concentration of the acid and its conjugate base, respectively. Buffering capacity is at its maximum when pH = pKa, and it ranges between pH = pKa ± 1.

5 Measurement of pH of Milk and Whey Sample

5.1 Principle

As described in Subheading 1.

5.2 Materials and Reagents

1. Milk and whey samples.
2. Standard buffers (pH 4.0, 7.0, and 9.0).
3. pH meter.

5.3 Guidelines to be Followed While Operating the pH/Ion Meter

1. The instrument must be maintained when not in use and operated by following the manufacturer's guidelines.
2. The pH meter should be standardized before use, using the standard buffer.
3. Ensuring proper washing of the electrodes before and after use, avoid touching the electrode.
4. Make sure that the electrode should not get dried. As drying of the electrode damages the membrane.

5. Rubbing the electrodes with dry clothes or cleaning in an ultrasonic bath should be avoided.
6. Presence of any air bubbles in the membrane should be removed by gentle tapping of the electrode before use.
7. For removal of the denatured protein from the surface of the electrode, it may be dipped in a solution of dilute HCl and pepsin.
8. Temperature should be maintained strictly during analysis and calibration.
9. The level of the sample or the storage solution should not be more than that of the internal solution in the electrodes.

5.4 Standardization/Calibration of pH Meter

1. The standardization of the pH meter should be done as per the manufacturer's instruction.
2. Take the standard buffer of pH 4.0 and dip the electrode in it, swirl and wait for a stable reading, and then adjust the pH to 4.0.
3. The calibration should be done using at least two different buffers. Their pH should differ by 2–3 units.
4. Generally, the buffers of pH 4, 7, or 9 are used for calibration. Always use fresh buffers for calibration.
5. The used buffer(s) should be discarded.
6. The pH electrode must be washed with distilled water and wiped with tissue paper to remove the droplets of water or buffer or sample.
7. Similarly, calibrate the pH meter with other buffers at different pH and adjust the variation in the pH if any.

5.5 Measuring pH of the Sample

The sample, that is, milk, whey, or curd is taken in the beaker. The diaphragm of the electrode is dipped in the sample to measure its pH.

6 Applicability of ISE for Measuring the Ionic Calcium in Skim Milk

6.1 Principle

Activity and concentration of an ion must be considered using ISEs. Concentration is the amount of the ion present in all possible forms, that is, in bound and unbound or free form in the solution. Conversely, activity gives an idea of the chemical reactivity of an ion. In the solution, the ions interact either with the solvent or each other, as a result the effective/actual activity or concentration comes out to be lower than the actual concentration present. Both the terms, concentration and activity are related as follows:

$$A = f\,C,$$

where C is concentration, f is activity coefficient, and A is activity.

Activity coefficient depends on the ionic strength. For the adherence to the Nernst equation, it is important that the calibration liquid and the samples must be adjusted to a high but a constant ionic strength.

The prerequisites to be followed while using an ISE:

1. The reference potential during the measurement should be constant.
2. Eliminate interferences due to method and electrode.
3. Adjust ionic strength and the pH.
4. Temperature during analysis should be constant.

6.2 Preparation of Calibration Curve

Preparation of calibration curve is a must when working with ISE. Both the indicator and reference electrodes are dipped in the solution of known concentration and the electrode potential (mV) is recorded and plotted against the log of the concentration. In case of the test sample, the concentration of an ion is determined by using the recorded electrode potential of the sample and the calibration curve.

6.3 Materials and Reagent

1. Calcium chloride.
2. Potassium chloride.
3. Calcium ISE.
4. Reference electrode.
5. Ion meter.

6.4 Procedure

1. Prepare calcium chloride solutions having Ca^{2+} concentration ranging from 0.1 M to 10^{-4} M, 0.1 M KCl is also added to the solution in order to maintain a high ionic strength. Immerse the electrodes in the calcium solution in the order of their increasing concentration and record the corresponding output electrode signal in mV.
2. Construct a calibration curve between the log of calcium concentration and the measured electrode potential, mV.
3. Measure the calcium concentration in the milk sample and record the output signal from the electrode, that is, mV and calculate the concentration of Ca^{2+} ions from the calibration curve.
4. Multiple addition method discussed below can be used to correct the chelating effect in milk.
5. To 10 ml of milk sample taken in the beaker, aliquots of 100 μl of 0.1 M $CaCl_2$ should be added to a constant electrode response. This constant response is generally obtained by

adding four aliquots. Thus, the calcium ion (Ca^{2+}) concentration can be calculated by the Nernst equation, mentioned below

$$X = (A * V)/((V + Va) * B) - Va)$$

where

$B = 10^{(E-E_a)/S}$

A and X are the Ca^{2+} ion concentration (mmol) in the standard and milk sample, respectively.

V and Va are the quantity of standard and sample in ml, respectively.

E_a is the potential (mV) measured by the electrode before adding the standard.

E is the potential (mV) measured by the electrode after adding the standard.

S is electrode sensitivity, that is, the change in electrode potential (mV) when the Ca^{2+} ion activity is altered by a factor of 10.

References

1. Orellana G, Cano-Raya C, López-Gejo J, Santos AR (2011) 3.10—online monitoring sensors. In: Wilderer P (ed) Treatise on water science. Elsevier, Oxford, pp 221–261. https://doi.org/10.1016/B978-0-444-53199-5.00059-2
2. https://www.chegg.com/homework-help/questions-and-answers/really-appreciate-some one-especially-explain-right-side-graphspecifically-laf3-solid-stat-q84924608
3. Harvey D (2013) Liquid membrane ion-selective electrode | image and video exchange ForumImage and video exchange forum. https://community.asdlib.org/imageandvideoexchangeforum/2013/07/31/liquid-membrane-ion-selective-electrode/. Published 31 July 2013
4. Nielsen SS (2017) Introduction to food analysis. Food analysis. Springer, pp 3–16

Check for updates

Chapter 8

Spectroscopy

Abstract

Spectroscopy involves the production, measurement, and interpretation of the spectra produced as a result of the interaction of the matter with electromagnetic radiation. It has emerged as a wonderful technique that can be used to solve a large number of analytical problems. In this chapter, basic concepts like electromagnetic spectrum, Planck's Quantum theory, and De Broglie concept have been covered. The principle of spectroscopy which is Beer–Lambert law along with its fundamental, chemical, and instrumental limitations has been also covered. The design of spectrophotometer along with the working of its components like radiation source, wavelength isolator, sample holding cell, photoelectric detector, and readout device has been included. Basic differences between the design of single and double beam spectrophotometer have been discussed. Topics of fluorescence spectroscopy have also been touched upon. Applications like determination of absorption spectrum of any analyte, verification of Beer–Lambert law, and effect of pH on the λ_{max}, absorbance (A), and absorptivity of p-Nitrophenol solution are also included.

Keywords Electromagnetic spectrum, Planck's Quantum theory, De Broglie concept, Beer–Lambert law, Fluorescence spectroscopy, Single and double beam spectrophotometer

Spectroscopy is a technique which involves production, measurement, and interpretation of the spectra produced as a result of the interaction of the matter with the electromagnetic radiation. Spectroscopy has emerged as a wonderful technique that can be used to solve large number of analytical problems. Both quantitative and qualitative spectroscopy techniques are frequently utilized for the analysis of composition of food. Qualitative measurements utilize the fact that the absorption or emission of an analyte depends on the energy spacings which are unique to it. Conversely, analyte concentration in the sample solution determines the absorption or fluorescence at a specific wavelength in quantitative measurements. Commonly used spectrochemical methods include infrared absorption, visible, ultraviolet spectroscopy, molecular fluorescence spectroscopy, and nuclear magnetic resonance spectroscopy [1–3]. Generally during the analysis of food, the radiation(s) in the ultraviolet and visible (UV-Vis) regions are frequently used.

Kamal Gandhi et al., *Advanced Analytical Techniques in Dairy Chemistry*, Springer Protocols Handbooks,
https://doi.org/10.1007/978-1-0716-1940-7_8,

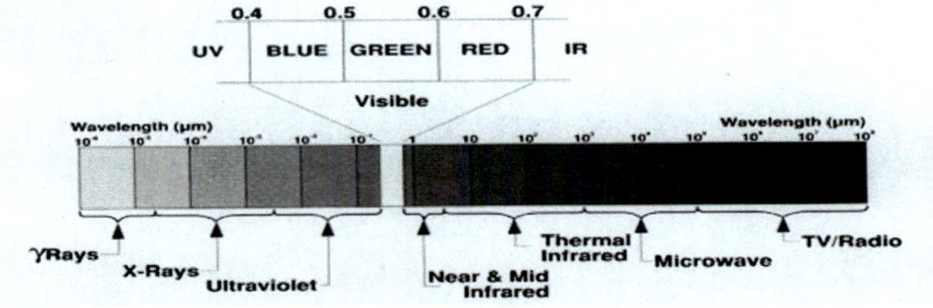

Fig. 1 Electromagnetic spectrum

Table 1
Wavelength limits of different electromagnetic radiations

Wavelength region	Wavelength limits	Type of transitions
Gamma ray	0.01–1 Å	Arrangements in protons/neutrons
X-ray	0.1–10 nm	Inner shell electrons
Ultraviolet	10–380 nm	Outer shell electrons
Visible	380–750 nm	Outer shell electrons
Infrared	0.75–1000 μm	Vibration position of the atoms
Microwave	0.1–100 cm	Rotation position in molecules
Radio wave	1–1000 m	Nuclei orientation in magnetic field

Wavelength ranging from approximately 200–700 nm falls under the UV-Vis portion of the electromagnetic spectra. The UV and visible radiation range from 200 to 350 and 350 to 700 nm, respectively. The UV light cannot be perceived by the human eye, as it is colorless unlike the wavelengths falling in the visible range which have a distinctive color, falling between violet (short wavelength) to red (long wavelength) end of the spectrum.

Electromagnetic Spectrum

When the electromagnetic radiations are arranged in the increasing or decreasing order of their frequency or wavelengths, it is termed as electromagnetic spectrum as follows (Fig. 1):

Table 1 represents the wavelength range of different electromagnetic radiations and the type of transitions they are involved in.

Planck's Quantum Theory

Emission of radiation from any source is not continuous but occurs in discrete packets or bundles. Each packet is termed as quantum and group of packets is termed as quanta. Energy associated with the quantum is given by

$$E = \mathrm{h}\nu$$

For quanta, expression will change to

$$E = nh\nu$$

where h is the Planck's constant (6.626×10^{-34} J.s), n is the number of quantum making up the quanta, and ν is the frequency.

De Broglie Concept (Dual Nature of Matter)

As per the de Broglie concept, every matter in this universe comprises two natures, that is, particle and wave nature. On the similar lines, light is also considered to have dual nature: wave like and particle like.

Consider a photon with energy.

$$E = hv = hc/\lambda \tag{1}$$

As per Einstein, energy associated with any particle is given by.

$$E = mc^2 \tag{2}$$

Comparing Eqs. 1 and 2, we get.

$$\text{Mass of photon } m = h/c\lambda$$

Thus, momentum of photon.

$$\begin{aligned} p &= mc = hc/c\lambda = h/\lambda \\ \text{Or} \quad \lambda &= h/p \end{aligned} \tag{3}$$

For a particle with mass "m" and velocity of "v," momentum will be given by $p = mv$ and wavelength of the wave associated with it is given by.

$$\begin{aligned} &h/mv \\ &\text{Or} \\ &\lambda = h/p \end{aligned} \tag{4}$$

λ is called de Broglie wavelength.

The wave theory of the electromagnetic radiation explains the phenomena like interference, diffraction, and refraction, while the particulate nature of light explains the interaction between matter and light, thus forming the basis of emission and absorption spectroscopy. Apart from light, basic fundamental units of matter like electrons, protons, and neutrons also exhibit wave like behavior.

1 Principle

When a beam of light (electromagnetic radiation) is made to pass through a solution or a sample, majority of the radiation gets transmitted without losing its intensity, but a part of it gets attenuated at a selected wavelength. This process of attenuation is also referred to as absorption. The attenuation of the electromagnetic radiation is explained by terms like transmittance and absorbance (Fig. 2) [3].

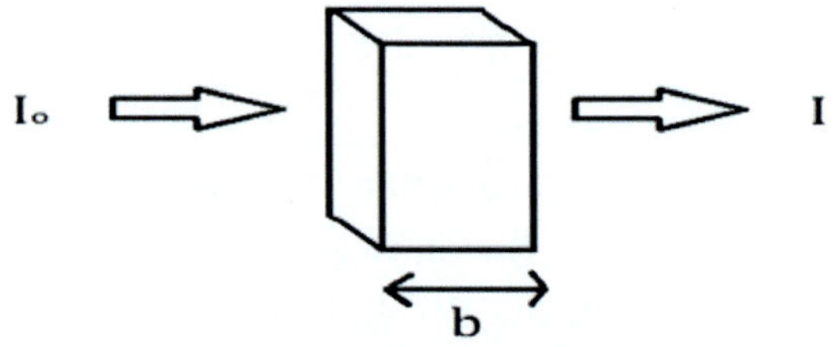

Fig. 2 Absorption of light by the analyte in the solution

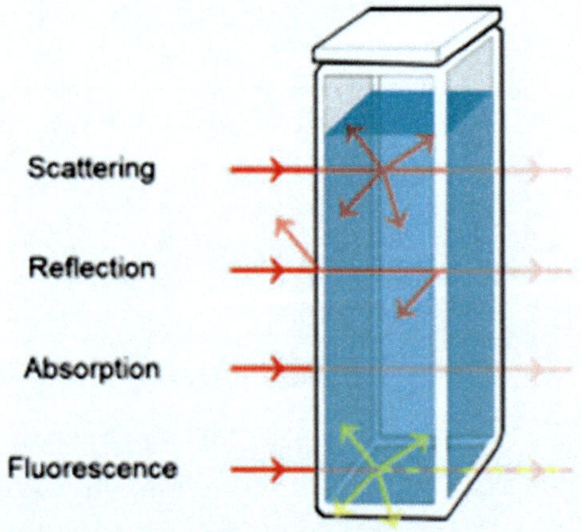

Fig. 3 Attenuation of the radiation on passing through a cuvette containing the absorbing solution (Source: Biochrom [4])

The ratio of the intensity of the electromagnetic radiation transmitted from the sample, I, to intensity of the incident light from the source on the sample, I_0 is termed as transmittance.

$$T = I/I_0 \tag{5}$$

Percent transmittance ($\%T$) varies between 0% (complete absorption) to 100% (no absorption). If the radiation passing through the sample gets attenuated its transmittance is less than 1. Other factors which cause the attenuation of the radiation include scattering of radiation, reflection from the sample container, and absorption of the radiation by the various components present in the sample other than the target analyte (Fig. 3).

This loss in the intensity of the radiation is compensated by using a blank. The radiation transmitted from the blank is recorded as I_0. Attenuation can also be expressed in terms of absorbance, A which is given by

$$A = -\log T = -\mathrm{Log}\frac{I}{I_0} = \mathrm{Log}\, I_0/I \tag{6}$$

or

$$\text{Log } I_0 - \log I = A \tag{7}$$

I_0 is always set to 100% and I is the $\%T$ read on the instrument, therefore

$$2 - \log \%T = A \tag{8}$$

As absorbance (or optical density) maintains linearity with the concentration of the analyte, so it is the most commonly used unit.

2 Beer's Law

Suppose, a radiation (monochromatic in nature) is made to pass through a sample having thickness dx, it will experience a drop in its intensity which is denoted by dI.

This decrease in intensity varies proportionately with the dx (Lambert's law) and the analyte's concentration (Beer's law), C; thus.

$$-\frac{\mathrm{d}I}{I} = \alpha C \mathrm{d}x \tag{9}$$

where I is the intensity of the radiation incident on the sample and a is the proportionality constant. Integrating Eq. 9 on both sides, where $x = b$ is the sample's overall thickness

$$\int_{I=I_0}^{I=I} \mathrm{d}I/I = \alpha C \int_{x=0}^{x=b} \mathrm{d}x \tag{10}$$

$$\ln \left(\frac{I_0}{I_t} \right) = \alpha b C \tag{11}$$

Using Eq. 6, Eq. 11 can be rewritten as:

$$A = abC \tag{12}$$

where "a" is the analyte's absorptivity (cm^{-1} conc^{-1}). Expressing the concentration of the sample in terms of molarity, molar absorptivity, ε (cm^{-1} M^{-1}) is used instead of absorptivity

$$A = \varepsilon bC \tag{13}$$

The probability of an analyte to absorb a photon having a certain amount of energy is given by the molar absorptivity and absorptivity. The wavelength of electromagnetic radiation determines the values for "a" and ε.

3 Limitations to Beer's Law

The Beer–Lambert law states that the concentration of the chromophore (analyte) in the solution and the absorbance of the solution is directly related. Construction of a calibration curve between the absorbance and concentration of analyte is a straight line with the slope being *ab* or *εb* and its intercept being 0. However, under certain conditions the calibration curves are nonlinear. The deviations from the linearity are categorized as fundamental (due to the limitations of the law), chemical (due to the nature of the chemical species under analysis in the sample), and instrumental (deviations occurring due to the instrument) [5].

3.1 Fundamental Limitations

Beer's law stands true for dilute solutions having a concentration of about 10 mM. The concentration and the chemistry of the analyte in the solution to be analyzed become the limiting factor. Increasing the concentration of the analyte results in increase in the intermolecular interaction with a subsequent decrease in the intermolecular distances. This disturbs the charge distribution of the molecules as they interact with each other. These effects alter the photon capturing ability of the analyte at a specific wavelength, thus altering the absorbance of the analyte. Another cause for deviation is that both "*a*" and "*ε*" depend on the sample's refractive index, which alters with the analyte's concentration. RI for dilute samples remains essentially constant, hence obeying the law.

3.2 Chemical Limitations

When the absorbing species get involved in various equilibrium reactions like the association or dissociation of the analyte molecules or ionization of the weak acid in a non-buffered solution. This leads to the alteration in the predominant form of the analyte when its concentration is varied. The absorptivity "*a*" of the ionized form of the analyte versus its neutral form will be different, thus disturbing the linearity between the concentration and the absorbance.

3.3 Instrumental Limitations

The two major limitations in Beer's law arising due to the instrument are that the law stands true if and only if the radiation or the light source passing from the solution is monochromatic. In case it is polychromatic, it will produce different values of the absorptivity for different wavelengths and in that case the law is not obeyed. The stray light is contributed by the spectrophotometer, though received by the detector, but it is not taken into consideration in the spectral band which has been isolated by the monochromator. Thus, the intensity of the selected wavelength of light should be 10 times greater than that of the stray light, so as to obtain a reasonable signal-to-noise ratio. If the stray light gains intensity, the absorbance measured by the detector will be independent of the chromophore or the analyte concentration.

4 Factors Influencing the Absorption Spectra of Chromophores

Different types of shifts occurring in the absorption spectra of a chromophore from its standard position are classified as:

Bathochromic (red) shift, hypsochromic (blue) shift, hyperchromic shift, and hypochromic shift. The red and blue shifts occur when the λ_{max} gets shifted towards the longer and shorter wavelength, respectively, while the other two shifts, that is, hyperchromic and hypochromic shift occur when the absorbance, λ_{max} or $\varepsilon_{\lambda_{max}}$, gets shifted towards a greater intensity and lesser intensity, respectively.

These shifts occur if auxochrome gets conjugated with a chromophore or any variation is there in its pH, ionization state, or polarity of the solvent. The width of the slit of the instrument also affects the resolution of the two closely spaced peaks in the absorption spectra [5].

5 Energy Levels in Atoms and Molecules

The difference in the energy of the molecule or atom in its energy state and in the ground state is called as the relative potential energy of a molecule or atom. Figure 4 depicts the potential energy level of an organic molecule in which the bottom line (bolded) represents the ground state or the lowest energy state. The energy states are classified into three states each has its own rotational and vibrational energy levels. The potential energy of the electrons varies with the orbital level. Electronic transition (change in the energy levels) is

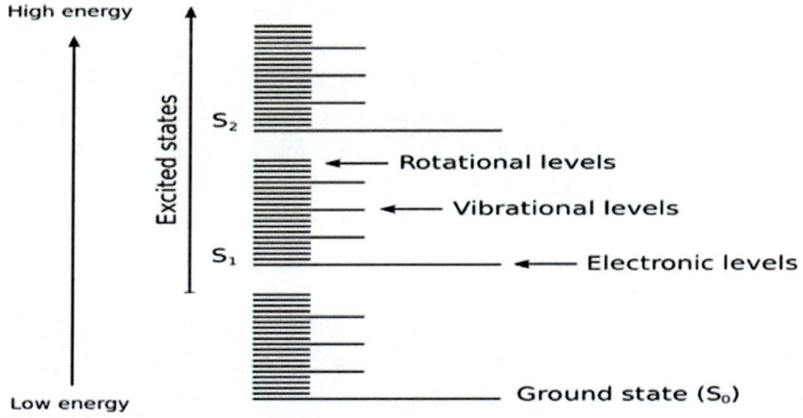

Fig. 4 Molecular energy level diagram of a molecule

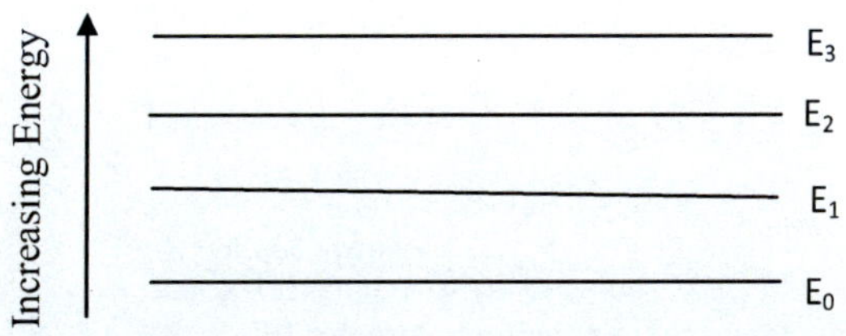

Fig. 5 Energy level diagram of an atom

termed as the situation in which an electron absorbs or emits a photon of appropriate energy on changing its orbital. The potential energy of the whole atom/molecule gets affected with any increase or decrease in the potential energy of an electron. Thus, an energy level diagram of an atom (Fig. 5) consists of only electronic energy levels. Conversely, these energy levels of atoms are devoid of their corresponding rotational and vibrational levels, thus appearing less complicated than the molecules. The magnitude of the energy difference of the valence electrons in the atom is given by the energy difference between the first excited state and its ground state. This energy is of the same magnitude as the energy associated with the photons of UV-Visible radiation. Vibration levels lie within the electronic energy levels which in turn covers the rotational energy levels. To conclude, the internal energy of an atom is equivalent to the energy associated with the electronic energy levels and that for the molecule equals the energy associated with the rotational, vibrational, and electronic energies [6]. The equation is given as:

$$E_{\text{atom}} = E_{\text{electronic}}$$

$$E_{\text{molecule}} = E_{\text{electronic}} + E_{\text{vibrational}} + E_{\text{rotational}}$$

6 Components of UV-Vis Spectrophotometer

Spectrometers are the instruments which employ the use of monochromators for selection of the wavelength. When the transmittance from the instrument is the ratio of the two radiant powers, the instrument is referred to as a spectrophotometer. The main components of a basic spectrophotometer are: the monochromator, light source, the radiation detector, a readout device, and the sample/reference holder. A power supply is a must for the operation of the instrument. Figure 6 shows the basic design of a spectrophotometer.

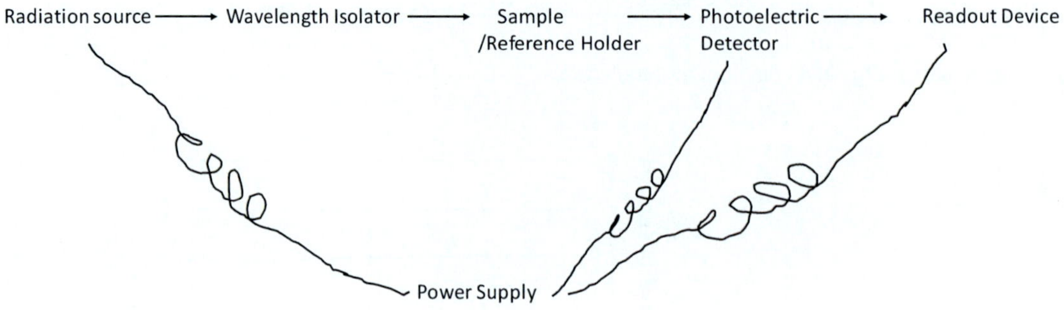

Fig. 6 Design of single-beam spectrophotometer [6]

6.1 Sources of Electromagnetic Radiation

A source of electromagnetic radiation must generate a stable and intense output for the region for which it is designed. The emitted radiation must have sufficient intensity so as to generate an adequate detector response and should not get affected with any change in the wavelength or drift significantly over a period of time.

Tungsten filament lamp is a widely used radiation source for Vis spectrophotometers, while these lamps are also used in NIR. Deuterium electrical-discharge lamps are most commonly used as a radiation source in the range of 160–375 nm (UV range). They are equipped with quartz windows and should be used with quartz sample holders as glass is found to absorb radiations below 350 nm significantly. Sources of electromagnetic radiation are grouped into two classes, namely continuum and line sources. The former emits electromagnetic radiation at a variety of wavelengths with variance in intensity as a function of wavelength, whereas the latter emits radiation at a narrow wavelength range.

6.1.1 Wavelength Selector

An ideal wavelength selector should have a high throughput of radiation with a narrow effective bandwidth. High throughput means that a higher number of photons should pass through the wavelength selector, which leads to the generation of a stronger signal with a subsequent lesser background noise, while a narrow effective bandwidth provides higher resolution. Spectrophotometers are equipped with various devices as a wavelength isolator. This is achieved by using an interference or absorption filter. It absorbs a selective radiation from a narrow region of the electromagnetic spectrum. These filters have a limitation that a continuous selection of the wavelength is not possible. This problem can be addressed by the use of monochromator, which converts a polychromatic radiation into the monochromatic radiation. A monochromator consists of two types of slits, that is, one at the entrance and another at the exit, concave mirror(s), and a dispersing element. The slit at the entrance makes way for the polychromatic light to enter the monochromator and gets culminated by the

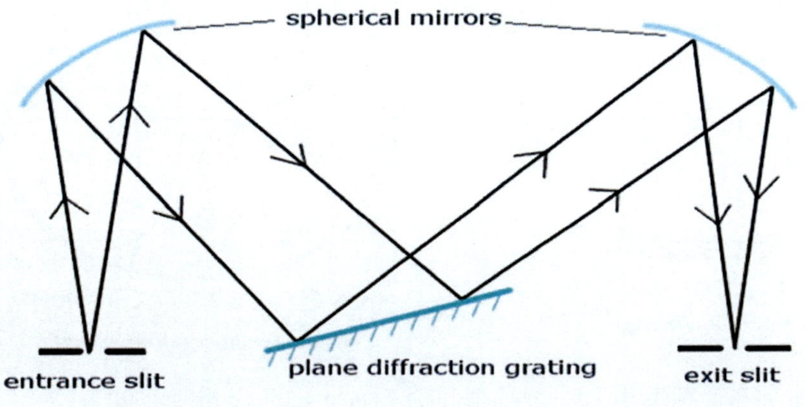

Fig. 7 Design of a monochromator [7]

concave mirror. This culminated polychromatic radiation is then dispersed into several radiations of different wavelengths. Nowadays, gratings serve the purpose of a monochromator. They are optically reflecting surface containing a large number of parallel grooves. The radiation falling on the grating gets dispersed which gets reflected by a second mirror on a plane surface consisting of an exit slit (Fig. 7). In some cases, the diffraction grating is replaced by a prism. The radiation of different wavelengths is then reflected from a concave mirror which focuses it sequentially along the focal plane. The radiation which lines itself with the exit slit of the focal plane is then emitted from the monochromator (Fig. 7).

6.2 Sample Holding Cell

The sample holding cell or cuvette for the spectrophotometric measurement should be selected according to the spectral region to be used. They have varying compositions and dimensions. The material used for the sample holding cell should not absorb any radiation in the spectral region selected for measurement. Cells which are made up of quartz/fused silica are used for the measurement in UV range, while for visible range, silicate glass is used. In some cases, plastic cells are also used in the Vis range. The dimension of the cell plays an important role with regard to the amount of the solution required and the path length during measurement. The approximate dimension of a typical absorption cell is 1 cm^2 and 4.5 cm long. For measurements, the minimum volume of the sample and the path length is 1.5 ml and 1 cm, respectively. Commercially available cells have a path length ranging from 1 to 100 mm. Narrow cells having path lengths of 1 cm and 4 mm in width are also available. Such types of cells are used when the amount of the sample solution is less than 1 ml [6].

6.3 Detectors

Currently various types of detectors are available but the two most popular of them are the phototube and the photomultiplier tube. They convert the energy of the incoming photons into the

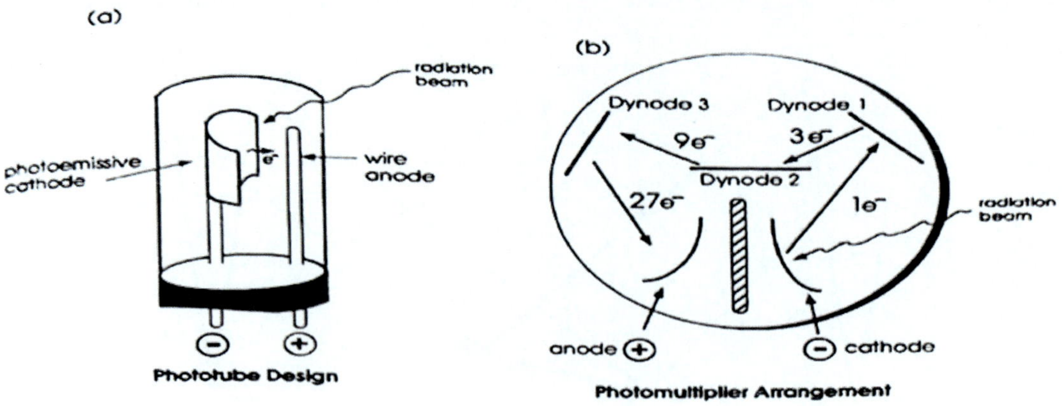

Fig. 8 Design of (a) Photodiode (b) Photomultiplier tube (Source: Suzanne Nielsen [6])

corresponding electrical current. The phototube (Fig. 8a) consists of a semi-cylindrical cathode with a photoemissive surface and a wire anode which are housed in a transparent tube under vacuum. The photons strike the photoemissive surface of the cathode, resulting in the release of the electrons, which travel towards the anode and get collected on it. This creates a potential difference leading to the generation of the current. The current generated by the virtue of the electrons emitted from the cathode is directly related to the no of photons generated or the radiant power of the beam, thus generating a current. The emission of electrons and hence the current produced are directly related to the number of photons, or radiant power of the beam, striking the photoemissive surface. The photomultiplier tube (Fig. 8b) is of similar design but it has an amplificating system which amplifies the number of the electrons getting collected at the anode for every electron which strikes the photoemissive surface of the cathode. The dynode (positively charged) attracts the electrons emitted from the cathode surface. These electrons then lead to the emission of more number of electrons, thus causing signal amplification. The signal amplification continues as the photomultiplier tube consists of a number of dynodes, each dynode capable of amplifying each electron. This will continue until the electrons emitted from the last dynode will get collected by the anode of the photomultiplier tube. The final number of electrons collected can be as much as 10^6 to 10^9 electrons collected per photon [6].

Currently, the detectors are equipped with a transducer which converts the signal consisting of photons into its corresponding electrical signal. Ideally, a linear relation should exist between the detector's signal and power of the electromagnetic radiation.

$$S = kP + D$$

where D is the detector's dark current (background current when no radiation is emitting from the source) and k is the detector's sensitivity.

6.4 Signal Processors

The detector output signal is enhanced and shown in a form that the user can clearly understand. Complexity of the system will decide the final form of the displayed signal. Analog meter is the simplest form of the analog signal generated from the detector in which the position of a needle is calibrated in percent transmission and/or absorbance.

For routine purpose, the analog signals are most appropriate, while the analog meters are difficult to read, thus resulting in generation of the data with a lower precision than that obtained from a digital meter or device. Digital readouts provide the signal as digits on the face of a meter. In these cases, processing of the signal is done between the analog output from the detector and the final digital display. In all the cases, the signal processor displays the final signal in the form of transmittance or absorbance. Majority of these instruments are equipped with microprocessors which are capable of extensive data manipulation on the digitized signal. For example, the final results from some spectrophotometers are displayed in concentration units, provided if proper calibration of the instrument is carried out with reference standards of appropriate concentration.

A device like a meter or computer displays the signal obtained from the transducer in readable form. The signal produced by the transducer is transmitted via a signal processor and seen on an analyst's computer. Calibration of the detector's response, amplification of the signal from the detector, removal of noise by the means of filtering, or mathematical transformation of the signal are accomplished by signal processor.

7 Single-Beam vs Double-Beam Spectrophotometer

The optical systems of spectrophotometers are categorized as either single-beam or double-beam instruments. In the former, the incident or the radiant beam travels only on a certain single path, that is, from the source through the sample and finally to the detector (Fig. 6). In these types of instrument, the transmittance is first set to 100% using a reference sample or blank before analyzing the sample. The blank and the sample are read one by one as there is only a single light path from the source through sample holding cell to the detector. Figure 9 shows the design of a double-beam spectrophotometer in which the beam is split into two radiations: one of which passes through the sample cell and other through the reference cell. The beam passes alternatively through the reference and the sample cell by the means of a chopper with alternate reflective and transparent sectors. The double-beam design enables

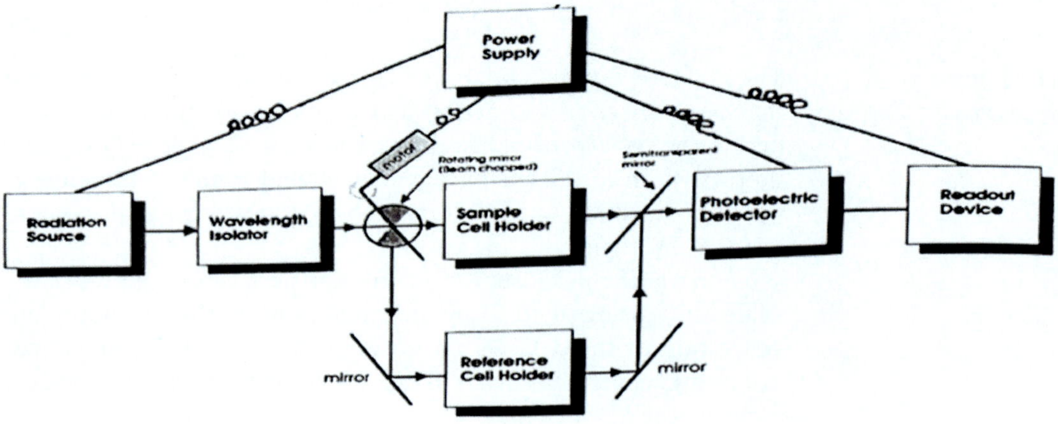

Fig. 9 Design of a double-beam spectrophotometer (Source: Suzanne Nielsen [6])

the measurement and comparison of the relative absorbance between the sample and reference cell. These types of instruments have an added advantage that the deviations or drifts in the radiant output from the source get compensated as the sample and reference cells are compared many times per second. Since the incident radiant beam is split, the power of the beam gets diminished. As the beam is split, the double-beam design generally provides inferior signal-to-noise ratio [6].

8 Fluorescence Spectroscopy

The fluorescence spectroscopy procedure is typically one to three orders more sensitive than corresponding methods of absorption spectroscopy. The measured signal is of the electromagnetic radiation produced by the analyte as it settles down to its corresponding ground state from the excited electronic energy state. The analyte when absorbs radiation in the UV or visible region gets excited to a higher energy level and gets activated. This activation and deactivation occur simultaneously during the measurement. Every molecular system has a distinct wavelength optimum for excitation and another distinct wavelength optimum for fluorescence emission. These wavelengths depend on the chemical makeup of the sample or system under study. The instrumentation (Fig. 10) of a fluorescence spectrometer is same as that of the UV-Vis absorption spectrometer. But there are definite differences between the two in terms of arrangement of the optical system. Wavelength selectors are not required in these instruments for excitation beam and emission beam. In some simple fluorimeters, both wavelength selectors are filters such that the excitation and emission wavelengths are fixed.

Grating monochromators are used for selecting excitation and emission wavelengths in more sophisticated spectrofluorometers. In fluorescence spectroscopy, the source of radiation and the detector is arranged at right angles to each other, thus minimizing the

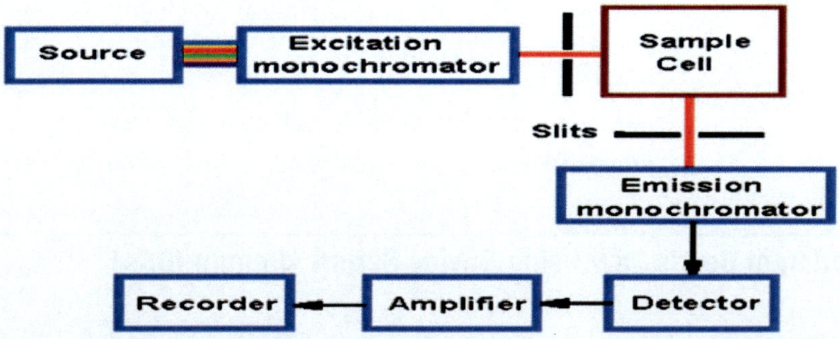

Fig. 10 Design of fluorescence spectroscopy [8]

interference between the transmitted radiation and that scattered from the sample. The change in the radiant power of the source beam emitted as it passes through the sample is proportional to radiant power of the fluorescence (P_F) light. One of the drawbacks of fluorescence spectroscopy is that only those molecules which show fluorescence can be determined using it.

9 Determination of Absorption Spectrum of Bovine Serum Albumin (BSA)

9.1 Principle

Choosing the appropriate wavelength to carry out measurements is important. The wavelength chosen should be such that analyte under study shows maximum absorbance and there is no rapid change in the absorbance. λ_{max} is the wavelength at which an analyte shows the maximum absorbance. Measurement should be made at this wavelength as it increases the sensitivity, that is, change in the absorbance with per unit change in the analyte's concentration will be higher and also there is greater adherence to the Beer–Lambert law as molar absorbtivities of the analyte will not be affected due to the narrow range of wavelength that is being used.

9.2 Materials and Reagents

1. Test tubes.
2. 0.5 mg/ml solution of BSA.
3. Spectrophotometer.
4. Tissue paper/filter paper.
5. Quartz cuvette.

9.3 Procedure [5]

1. Switch on the spectrophotometer and give a warm-up time for 10–15 min.
2. Wavelength is then adjusted to 220 nm.
3. Measure the transmittance for distilled water and adjust it to 100.
4. Scan the BSA solution between the wavelength range from 220 to 620 nm in the other cuvette and measure their absorbance.
5. In case, if the range of wavelength at which the absorbance is measured is high, the absorbance should be measured at an interval of 2 nm, so as to identify λ_{max}.
6. Draw a graph between absorbance and wavelength and determine the λ_{max}.

10 Verification of Beer's Law Using Bovine Serum Albumin (BSA)

10.1 Principle

As mentioned in Subheading 1.

10.2 Materials and Reagents

1. Test tubes.
2. 0.5 mg/ml solution of BSA.
3. Spectrophotometer.
4. Tissue paper/filter paper.
5. Quartz cuvette.

10.3 Procedure

1. Prepare BSA solution of varying concentration (0.05–0.1 mg / ml).
2. Determine the absorbance of the prepared solutions at the obtained λ_{max} as in Sect. 9.3.
3. Calibration curve between concentration and absorbance is then determined and its linearity is checked.

   Observations: The Beer law is obeyed for the range during which graph is linear.

11 Effect of pH on the λ_{max}, Absorbance (*A*) and Absorbtivity of p-Nitrophenol Solution [5]

11.1 Principle

The ionization state of the ionizable chromophores depends on the pH of the solvent. Any change in the pH may lead to bathochromic or a hyperchromic shift in the spectra. This can be studied by subjecting the p-nitrophenol to different pH and determining the spectra.

11.2 Materials and Reagents

1. 1 M HCl and 1 M NaOH solutions.
2. Tissue paper/filter paper.
3. Test tubes.
4. Spectrophotometer.
5. p-Nitrophenol.
6. Quartz cuvette.

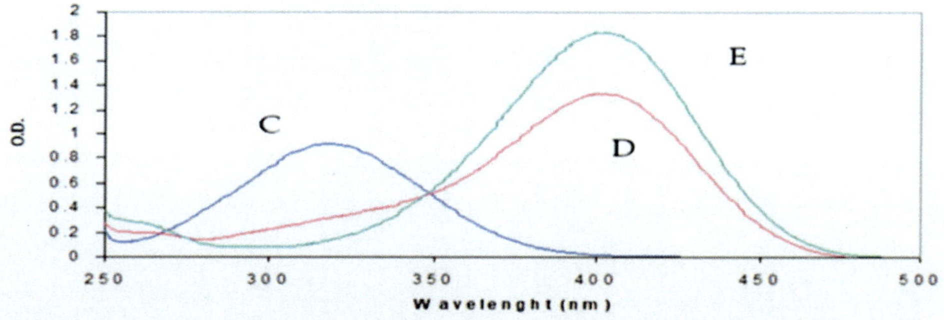

Fig. 11 Absorbance curves of the solutions at different pH in the range of 250–500 nm

Table 2
λ_{max}, absorbance, and molar absorbtivity of p-nitrophenol solution at different pH

p-Nitrophenol solution	pH	λ_{max}	Concentration (mM)	A at λ_{max}	ε at λ_{max}
Curve C (Fig. 11)	5	320	0.1	0.911	9110
Curve D (Fig. 11)	7	402		1.33	13,300
Curve E (Fig. 11)	9			1.83	18,300

11.3 Procedure

1. Adjust the pH of 0.1 mM solution of p-nitrophenol in distilled water to 5, 7, and 9 using 1 M HCl or NaOH.
2. Record their absorption in the range of 280–500 nm (Fig. 11) and draw the graph between the wavelength and absorbances and note down the value of λ_{max}.

11.4 Observations and Calculations

1. Determine λ_{max}, A and ε at λ_{max} and compare the results with Table 2.
2. Observed values will be constant at a particular pH and will change with change in the pH of the solutions.

References

1. Christian GD (2007) Analytical chemistry. Wiley, Hoboken, NJ. https://www.chromacademy.com/
2. Pearson D (1976) The chemical analysis of foods. Longman Group
3. Wilson K, Walker J (2010) Principles and techniques of biochemistry and molecular biology. Cambridge University Press
4. Evans L (2004) UV-VIS spectrophotometry: a brief background to spectrophotometry. Biochrom Ltd a Division of Harvard Bioscience, Inc, pp 1–15
5. Mann B, Sangwan R, Kumar R (2008) Research techniques in dairy chemistry. NDRI, Karnal
6. Nielsen SS (2017) Introduction to food analysis. Food analysis. Springer, pp 3–16
7. https://www.azooptics.com/Article.aspx?ArticleID=380
8. http://www.pci.tu-bs.de/aggericke/PC4/Kap_I/mol-fluo.htm

Check for updates

Chapter 9

Infrared (IR) Spectroscopy

Abstract

Infrared (IR) spectroscopy involves the measurement of diverse frequencies of infrared radiations through certain matrices like foods or other solids, liquids, and gases which are placed in the path of an infrared beam. IR spectroscopy mainly helps to find out the type of the functional group(s) which make up the sample. The frequency of the absorbed infrared radiation is unique or characteristic of every functional group. In this chapter, the basic principle, working, and components of IR spectroscopy have been covered. Design and efficiencies of the dispersive and Fourier transform instruments have been compared. Sample preparation and handling technique for use in FTIR spectrophotometer have been covered. A step-by-step procedure to interpret FTIR spectra has been provided. Both qualitative and quantitative applications have been discussed. Various types of ATR crystals materials have been compared. The basic principle and design of attenuated total reflectance spectroscopy has been discussed. Application of chemometrics on the FTIR spectral data for determining the quality of milk and milk products has been touched upon.

Keywords Vibrations, Functional groups, Chemometrics, Dispersive instruments, Fourier transform instruments, Attenuated total reflectance, Fingerprint region, Functional group region

Infrared (IR) spectroscopy involves the measurement of diverse frequencies of infrared radiations through certain matrices like foods or other solids, liquids, and gases which are placed in the path of an infrared beam. IR spectroscopy mainly helps to find out the type of the functional group(s) which make up the sample. The frequency of the absorbed infrared radiation is unique or characteristic for every functional group. Thus, IR spectroscopy plays a significant role in elucidating the structure of a compound along with its identification and quantification.

1 Infrared Region of the Electromagnetic Spectrum

The infrared radiation(s) fall next to the visible light in the electromagnetic spectrum. They generally have wavelengths (λ) shorter than the microwaves but longer than visible light (Fig. 1). Their wavelength ranges from a lower limit of 0.76–0.8 μm to an upper limit of 1 mm.

Kamal Gandhi et al., *Advanced Analytical Techniques in Dairy Chemistry*, Springer Protocols Handbooks, https://doi.org/10.1007/978-1-0716-1940-7_9,

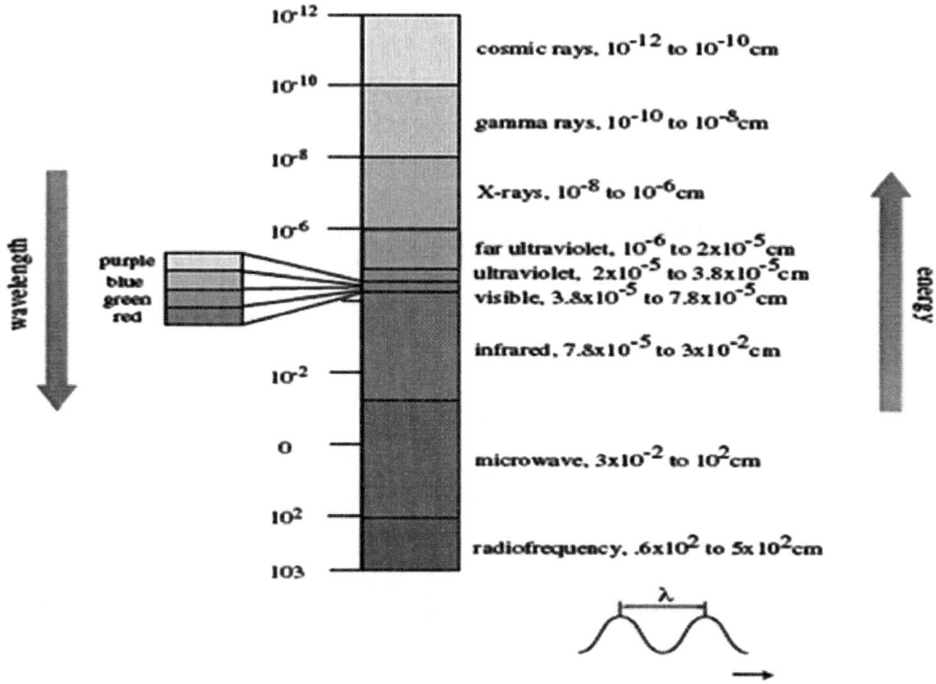

Fig. 1 Electromagnetic spectrum

Generally, in IR spectroscopy, the wavelengths employed for measurement range from 0.8 to 100 μm and are subdivided into three classes:

Near-IR (0.8–2.5 μm)

Mid-IR (2.5–15 μm)

Far-IR (15–100 μm)

Both near and mid-IR regions of the spectra are used for the qualitative and quantitative analysis of foods matrices. Apart from micrometer, frequency is also used as a unit to measure the IR. Generally, frequency is used because a direct relation has been established between energy and frequency of the radiation

$$E = h\nu$$

where $E =$ the energy of the system.

$h =$ Plank's constant, 6.6×10^{-34} J-s

$\nu =$ the frequency in hertz

Frequencies are also commonly represented as wavenumber ($= 1/\lambda$ in cm) $= 10^4/$ (λ in μm).

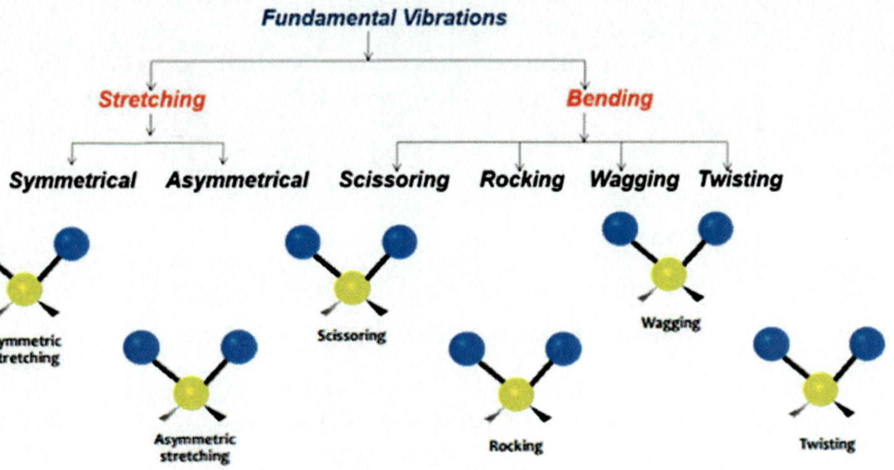

Fig. 2 Stretching and bending vibrations in a water molecule [5]

2 Molecular Vibrations

The atoms which make up the molecule are not stable. They keep on moving as if attached by a spring. The prerequisite for the molecule to absorb IR is if the vibration of the molecule is capable enough to change its charge distribution with a subsequent change in its dipole moment. There are a number of vibrations associated with the molecule but bending (scissoring) and stretching motions are the most significant type which can change or alter the dipole moment. The type of these vibrations is shown in a typical water molecule (Fig. 2).

For a linear molecule like CO_2, the number of vibrations is calculated by the formula 3 N – 5, where N is number of atoms, while for a nonlinear molecule it is calculated by the formula 3 N – 6. For a linear molecule like water, the frequencies for symmetric stretching, asymmetric stretching and scissoring (bending) vibrations are at 3652, 3756, and 1596 cm^{-1}, respectively. The values clearly indicate that higher energy is required by the stretching motions than the scissoring motions.

2.1 Infrared Activity

Diatomic homonuclear molecules like oxygen (O_2), hydrogen (H_2), nitrogen (N_2) possess no dipole moment (due to cancellation of the charges, Dipole Moment = Charge Distance), thus they do not absorb IR. The stretching of the bonds in such molecules will not produce a dipole moment and absorption of IR. Conversely, certain molecules which do not have a permanent dipole moment still absorb IR because the stretching of the bonds causes a change in their dipole moment. The common examples of such molecules are hydrogen chloride (HCl), carbon monoxide (CO) which are heteronuclear diatomic in nature.

To calculate the frequency of light absorbed, the following equation (Hooke's law) is used:

$$\nu_{\mathrm{osc}} = \frac{1}{2\pi}\sqrt{k\,\frac{m_1 + m_2}{m_1 m_2}} \tag{1}$$

where,

k = force constant which indicates the strength of the bond.

m_1 and m_2 are the masses of the two atoms.

Thus, Hooke's law clearly indicates that the strength of the bond is directly related to the force constant, for example, a C=C double bond in comparison to the C-C single bond will absorb a higher frequency of light. Also, the masses of the atoms are inversely related to the frequency of the light absorbed. When a molecule absorbs light, it undergoes transitions from lower energy state to a higher energy state. If the light absorbed by the atoms brings about changes in the rotational and vibrational energy states of the molecule (electronic energy > vibrational > rotational energy), the transitions give absorption in the IR region and the arising spectrum is called as IR spectrum [1, 7, 8].

The bonds of the atoms vibrate at certain frequencies which are determined by

- Atomic masses.
- Molecular shape.
- Bond stiffness.

Those frequencies of IR radiation which match with the vibrating frequency of bonds in molecules are absorbed [9]. Molecules of different compounds have unique combination of atoms. Each molecule absorbs specific wavelengths from IR radiations and different molecules do not produce exactly the same IR spectrum. Hence, IR spectrum of a sample can serve as a fingerprint. Peak size in spectrum is proportional to amount of the analyte molecule and is excellent tool for quantitative analysis.

3 Factors Affecting Absorption of Frequency

- Bond order: Higher the bond order, more will be frequency absorption because of more strength of the bond.
- Atomic mass: Lower is the atomic mass, higher will be the frequency absorption as evident from Eq. 1.
- Hydrogen bonding: More is the strength of the hydrogen bonding, higher will be the energy absorption.

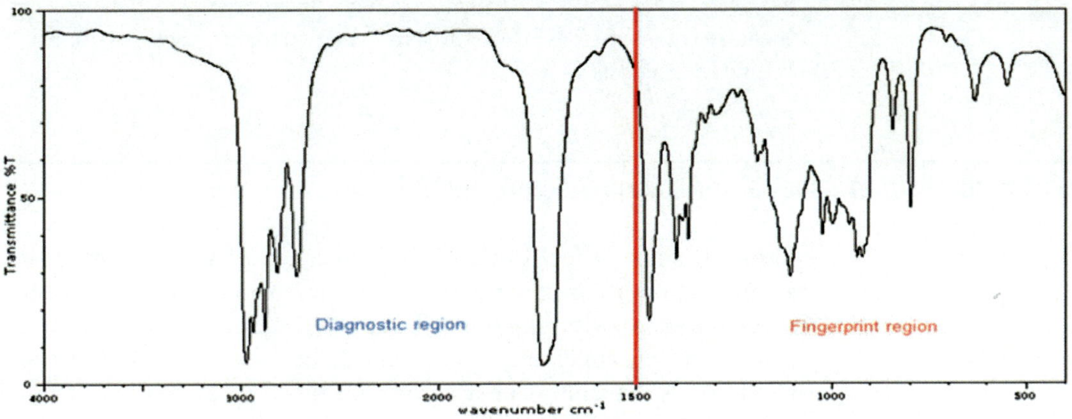

Fig. 3 IR spectra showing the fingerprint and functional region (11)

- Hybridization: SP^3 bond will absorb at a lower frequency as compared to SP^2 which is followed by SP bond.
- Resonance: Resonance weakens the double bond and absorption occurs at a lower frequency.

4 Regions of IR Spectra

The IR spectra have two regions: the functional group region and the fingerprint region (Fig. 3).

4.1 The Fingerprint Region

The spectra generated in the wave number between 600 and 1400 cm^{-1} contain the complex pattern of peaks, thus making it difficult rather impossible to assign a specific vibration to a specific group and thus is termed as the fingerprint region. No two compounds can have the same pattern of peaks in this region [7, 8]. Consequently, the identification of a compound can be done by referring to known spectra. Thus, the fingerprint region plays a substantial role in establishment of the identity of two compounds. Two different compounds, however, closely related cannot have identical or superimposable IR spectra. In fact superimposability of IR spectra of two compounds is taken as evidence for the identity of the two compounds.

4.2 Functional Group Region

The characteristic absorption band of functional groups is almost unchanged from one compound to another. The functional group region lies between 1400 and 4000 cm^{-1}. The infrared regions are subdivided into three well-defined regions, with every region furnishing distinct information [4]:

Far-infrared (400–33 cm^{-1}): vibrations of crystal lattice, vibrations of heavy atoms containing molecules, and vibrations of molecular skeleton.

Mid-infrared (4000–400 cm^{-1}): used in organic analysis.

Near-infrared (12,820–4000 cm^{-1}): overtones, used in quantitative analysis.

5 Comparison of Spectra of Alkanes, Alkene, and Alkyne

Figure 4 shows the IR spectra of n-hexane. As the alkanes have single bond in their structure between the carbon atoms, so the IR spectrum shows vibration contributed by the C–C and C–H bond. The vibrations appearing around 3000 cm^{-1} are attributed to the C–H bonds stretching and are most useful. Majority of the organic alkanes display such bands in their spectrum.

Figure 5 shows the IR spectra of 1-hexene. Apart from the presence of C–H bonds, alkenes also possess a double bond between the carbon atoms, which leads to the generation of the

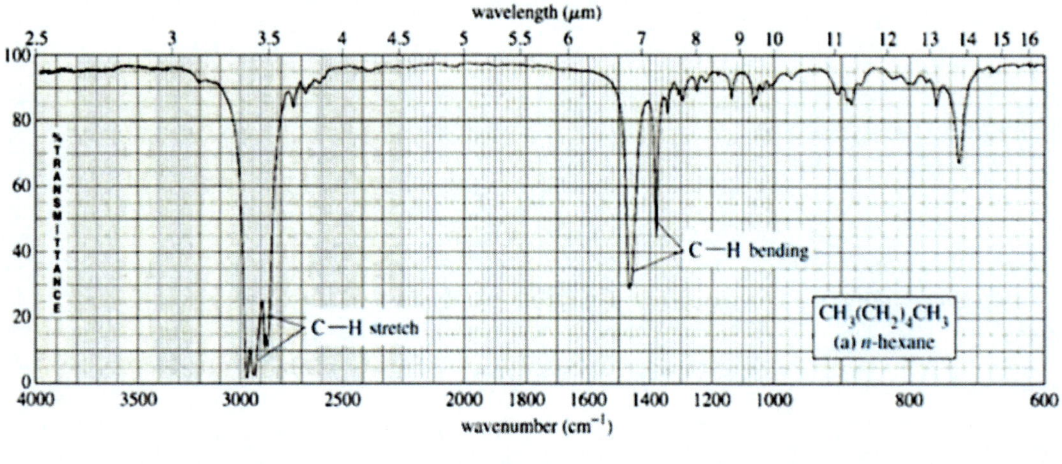

Fig. 4 Spectra of alkane [6]

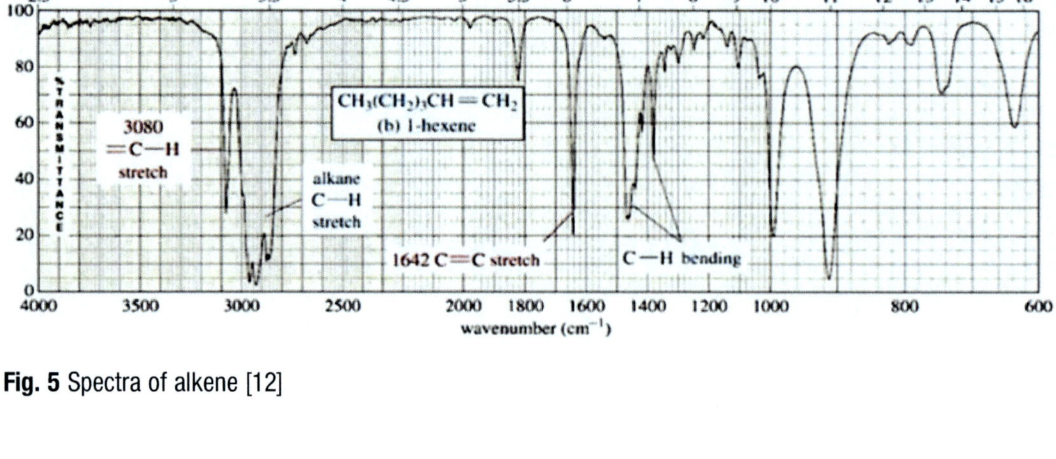

Fig. 5 Spectra of alkene [12]

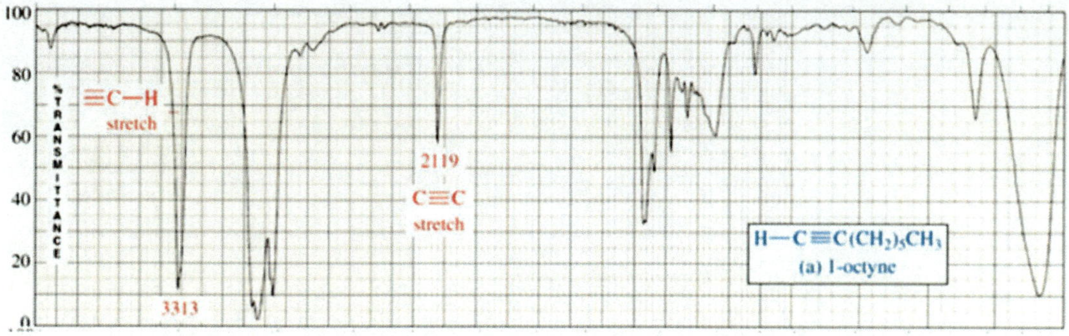

Fig. 6 Spectra of alkyne [6]

sharp, medium bands (due to the C=C bond stretching) around 1600–1700 cm^{-1}. Some of the alkenes may also generate a stretch at 3080 cm^{-1} due to the = C-H bond.

Figure 6 shows the IR spectra of 1-octyne. The triple bond between the carbon atoms generates the most significant band in alkynes. A sharp and weak band at about 2100 cm^{-1} is seen because the triple bond is not much polar in nature. The spectra generated by the highly symmetrical alkynes may not show such a weak and sharp band due to the lower polarity of the triple bond, while for terminal alkynes (having triple bond at the end of the carbon chain) which contain C-H bonds having the sp. hybridized carbon (the carbon atom involved in the triple bond) may show the weak and sharp band corresponding to the C-H stretching at about 3300 cm^{-1}.

6 Dispersive vs Fourier Transform Instruments

6.1 Dispersive Instruments

Dispersive instruments employ the use of a monochromator which disseminates the radiation of a particular frequency and allows them to pass through the sample. This allows the measurement of the amount of each frequency being absorbed by the sample (Fig. 7).

This type of IR spectroscopy is scarcely used to study food matrices because of the following reasons:

- Slow scanning process.
- Complex sample preparation and handling.
- Suitable data acquisition and processing systems are not available to provide quantitative information in a ready to use form.

Sources of IR spectrophotometers.

The most common IR source is made up of the ceramic core to which the coil of nichrome wire is wrapped, which glows on passing the electric current through it. Another type of IR source is made up of a silicon carbide rod generally called as a Globar, which can be

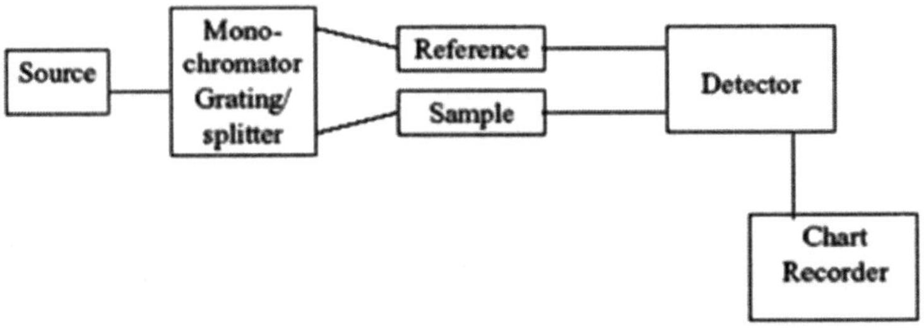

Fig. 7 Dispersive instruments

used as an intense source on applying a certain potential difference across it. Earlier, prisms made of NaCl were used as monochromator to disperse the radiation, but nowadays the NaCl prisms are replaced by diffraction gratings (as a monochromator). Thermocouple is the most commonly used detector, while currently more sensitive detectors like Golay detectors (a sealed tube filled with xenon) are being used. These detectors work when the radiation strikes the xenon gas which warms the gas leading to the change in the pressure within the tube. Semiconductor based detectors are those in which the conductivities of the semiconductors such as silicon or germanium vary with the amount of the radiation which strikes its surface.

A reference is used for two reasons:

- To avoid the fluctuation which might occur in the output, thus affecting the data.
- To cancel out the effect of the solvent (the reference represents the pure form of the solvent without the analyte).

6.2 Fourier Transform Infrared Spectroscopy

In Fourier transform (FT) instruments, radiation of all the wavelengths reaches the detector concurrently, that is, they are not dispersed. Typical IR spectrum is obtained after applying a mathematical calculation called as Fourier transform to the intensities of the light reaching the detector. The monochromator is replaced by a special instrument called as interferometer, which splits the radiation from the source into two beams. The advantage of FT instruments over dispersive instruments is that the spectra can be acquired more quickly, with more highly improved signal-to-noise ratio.

6.2.1 What is Fourier Transformation?

- Fourier transform is the computerized mathematical conversion of intensity vs. time spectrum into intensity vs. frequency spectrum. Fourier transformation can be expressed by the formula:

$$f(t) = \left(\frac{1}{N_k}\right) \sum_{\omega=-k}^{k} (F(\omega) \exp(i\omega t)$$

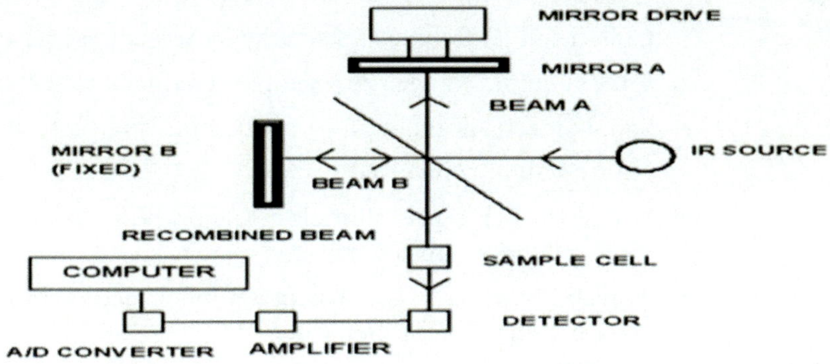

Fig. 8 Components of FTIR instruments

- Mathematical conversion consisting of sine and cosine wave functions is based on the fundamentals of quantum mechanics.

6.3 Components of FTIR Instruments

Michelson interferometer for FTIR spectroscopy comprises

- IR radiation source.
- Collimator.
- Beam splitter—half silver mirror.
- Fixed mirror.
- Moving mirror.
- He-Ne laser.

Figure 8 shows the components of FTIR instruments. IR radiations from source are collimated and directed to beam splitter.

6.3.1 Beam Splitter

Beam splitter divides IR beam into two beams. Half of the beam (50%) is reflected towards the fixed mirror, while the other half is transferred to the rotating or moving mirror. Both beams are reflected back and recombined.

Path length of beam is fixed for one and changing for other. A interference pattern is created.

Recombined beam passes through sample and reaches detector and produces "interferogram"—all IR frequencies "encoded" into it. Fourier transformation decodes IR spectrum. If the optical path difference (OPD) between the path of the light traveled between the stationary mirror and the beam splitter *vis-à-vis* that traveled between the moving mirror and the beam splitter is the integral multiple of wavelengths ($0, \lambda, 2\lambda, 3\lambda$, etc.), then we will have the constructive interference. If the OPD between the path of the light traveled between the stationary mirror and the beam splitter *vis-à-vis* that traveled between the moving mirror and the beam splitter is the half-integral multiple of wavelengths ($\lambda/2, 3\lambda/2, 5\lambda/2$, etc.), then we will have the destructive interference.

6.4 Advantages of FTIR in Food Analysis

- Extremely fast measurements—of the order of 1 s or so.
- Collects all IR frequencies simultaneously and scans at once.
- Time required to analyze a sample is reduced to few seconds.
- Simple mechanical design—only one moving part.
- No stray light is involved.
- No need of external calibration. Calibration can be done using He-Ne laser as internal standard.
- Availability of very convenient sampling accessories. Provides precise measurement method.
- Provides meaningful data of product at low cost and high speed.
- Small quantity of sample is enough to provide its composition.
- Versatile tool that can be used on almost any kind of sample, regardless of the physical state.
- Various sampling accessories are available which deals with a variety of samples to perform accurate analysis.
- Easy to operate and maintain.

6.5 Advantages of FTIR over Dispersive IR

Table 1 presents the differences between FTIR and Dispersive IR.

6.6 Sample Preparation and Handling Technique

Choice of sample preparation technique depends on:

- Type of information of the sample.
- Physical state of the sample (solid, liquid, gas).
- Limitation of the technique.

6.6.1 Gaseous Samples

Gaseous samples require little preparation beyond purification. The length of the cell column for the gasses should be long (typically 5–10 cm), as they have relatively weak absorbances.

6.6.2 Liquid Samples

Liquid samples are most commonly measured by transform IR spectroscopy, using cells having a path length of 0.01–1.0 mm. The sample to be analyzed is sandwiched or placed between the plates of a highly pure salt like NaCl or sometimes KBr or CaF_2 can also be used. Use of glass or quartz is prohibited as they have the ability to absorb in mid-IR region.

Demountable Liquid Cell

Fig. 9 shows the picture of a demountable liquid cell. It is suitable for qualitative analysis only. Types of samples which can be analyzed include nonvolatile liquid, nujol mulls, film samples, and paste. Since it is difficult to maintain the constant concentration and the amount of sample to be analyzed, hence this technique is not suitable for quantitative analysis.

Table 1
FTIR vs Dispersive IR

FTIR (Interferometer)	Dispersive IR
Only the mirror is in motion	A large number of moving parts
Produces a spectrum in as little as 1 s	Requires 7 ½ min to 30 min for one scan
Rapid scan speed	Slow scan speed
No slits in the system to define resolution	To improve resolution, the slits can be adjusted
Laser provides an internal calibration system with frequency accuracy	No internal reference for frequency accuracy
No equivalent to stray light	Stray light within the instrument
Sample further removed from the IR source	Thermal problems as the sample is close to the IR source
Emitted radiation by the sample is not viewed by the detector	IR radiation by sample is viewed by the detector
Improved signal-to-noise ratio	

Fig. 9 Demountable liquid cell [3]

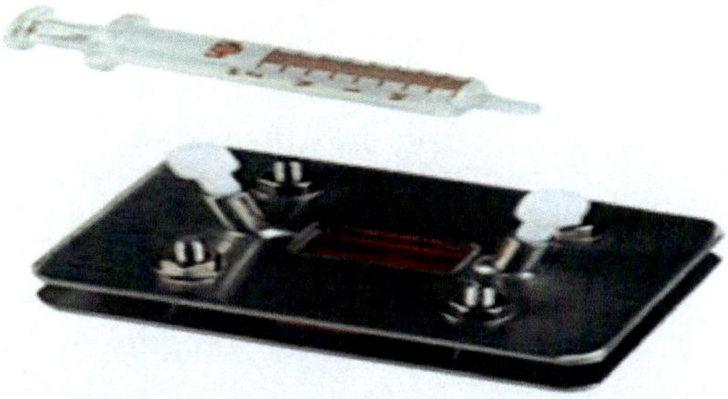

Fig. 10 Fixed thickness liquid cell [3]

Fixed Thickness Liquid Cell

Figure 10 shows the picture of a fixed thickness liquid cell. This technique is suitable for measuring the liquid or volatile samples quantitatively. Factory assembled with customer-specified cell thickness. There are three types of cell windows: NaCl, KBr,

Fig. 11 Preparation of KBr discs

KRS-5 (thallium bromoiodide) and their thickness lies between 0.025 and 5.0 mm.

6.6.3 Solid Samples

Solid samples can be prepared in four major ways:

1. The solid sample is crushed into fine pieces using a mulling agent (usually nujol-highly purified liquid paraffin) in a marble or agate mortar, with a pestle (Fig. 11). Then the crushed sample is applied on the salt plates as a thin film and measured.
2. The sample can also be prepared by grinding it finely (to prevent the scattering occurring due to large particles) with a highly purified salt like KBr. The powdery mixture is then converted to form a translucent pellet by crushing it in a mechanical die press. For preparation of the KBr discs, 2 mg of the sample is grinded along with 100–200 mg of KBr and then compressed into a transparent disc (Fig. 11). The evacuable die is employed to compress the powder mixture to form a cohesive disc 13 mm in diameter.
3. Cast film technique: The sample is first dissolved in a suitable, nonhygroscopic solvent (carbon tetrachloride, carbon disulfide, and chloroform). which is used mainly for polymeric materials. A drop of this solution is then positioned on to the surface of KBr or NaCl cell and the solvent is evaporated to leave a fine film of the sample on the cell which is analyzed directly. The film should not be too thick, as light cannot pass from it. This technique of sample preparation is used for the qualitative analysis of the polymeric materials.
4. In order to analyze failed plastic products microtomy is used, to cut a thin (20–100 micron) film from a solid sample. In this method, the integrity of the sample is preserved.

6.7 Interpretation of the Spectra

The vibrations occurring due to the stretching are generally considered while assigning a specific peak to a specific group and they are divided into four regions:

1400–650 cm^{-1} single bonds (other than hydrogen).

1900–1500 cm^{-1} double bonds.

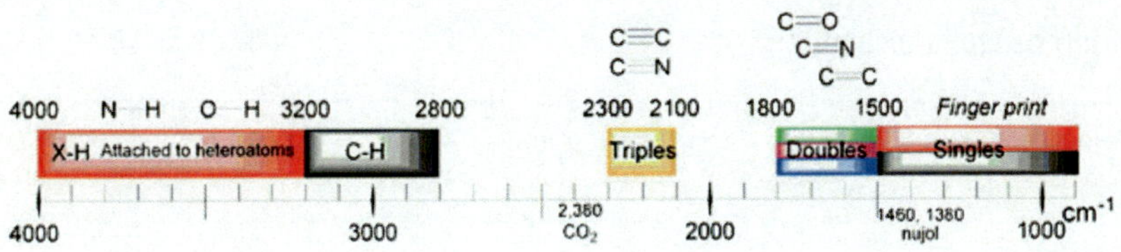

Fig. 12 Regions for different functional groups

2300–2000 cm^{-1} triple bonds.

3700–2500 cm^{-1} single bonds to hydrogen (Fig. 12 and Table 2).

The peaks in the wave number range 1650–650 cm^{-1} are due to bending vibrations but assigning a specific peak to a specific group is not possible.

Absorption of energy by bonds in organic molecule.

7 Applications of Mid-IR Spectroscopy

The spectra obtained are plotted with X-axis corresponding to the wave numbers or wavelengths, while on the y-axis transmittance or absorbance is plotted (Fig. 13).

7.1 Qualitative Applications

To identify a specific functional group present in an unknown substance, the central frequencies and relative intensities of the absorption bands can be used. Mid-IR spectrum of a substance can also be compared with a set of standard spectra and determining the closest match. Some of the food applications include the identification of flavor and aroma compounds using FTIR coupled with gas chromatography. IR spectra are also useful for the identification of packaging films.

7.2 Quantitative Applications

Infrared spectroscopic measurements obey Beer's law. Quantitative analysis involves determining the absorptivity. It is difficult to measure the absorptivity due to the low sensitivities of IR detectors, low intensities of IR sources, and the relative narrowness of mid-IR absorption bands. However, for quantitative measurements, empirical methods are used (Fig. 14).

Mostly this technique is used in the milk analyzers (Fourier transform infra spectroscopy) in which the major components of milk like fat, protein, and lactose are determined simultaneously. The wavelengths used are 5.73 μm (1745 cm^{-1}) for the CO groups of lipid, 6.46 μm (1547 cm^{-1}) for the NH_2 of protein, and 9.60 μm (1042 cm^{-1}) for the OH groups of lactose molecule. Automated version of these instruments can homogenize and analyze several hundred milk samples per hour. They can also be used to measure the degree of unsaturation in fats and oils.

Table 2
Regions for different types of bonds

Bond	Wavenumber/ cm^{-1}	Notes
Single bonds to hydrogen		
C–H	3000–2850	Saturated alkanes, limited value as most organic compounds contain C–H
=C–H	3100–3000	Unsaturated alkanes or aromatic
≡C–H	3300	Terminal alkyne
O=C–H	2800 and 2700	Aldehyde, two weak peaks
O–H O–H (free)	3400–3000 ~3600	Alcohols and phenols. If hydrogen bonding present peak will be broad 3000–2500 (e.g. carboxylic acids)
N–H	3450–3100	Amines: Primary—Several peaks, secondary—One peak, tertiary—No peaks
Double bonds		
C=O	1845–1800 and 1780–1740	Anhydrides
C=O	1815–1760	Acyl halides
C=O	1750–1715	Esters
C=O	1740–1680	Aldehydes
C=O	1725–1665	Ketones
C=O	1720–1670	Carboxylic acids
C=O	1690–1630	Amides
C=C	1675–1600	Often weak
C=N	1690–1630	Often difficult to assign
N=O	1560–1510 and 1370–1330	Nitro compounds
Triple bonds		
C ≡ C	2260–2120	Alkynes, bands are weak
C ≡ N	2260–2220	Nitriles
Single bonds (not to hydrogen)		
C–C	Variable	No diagnostic value
C–O, C–N	1400–1000	Difficult to assign
C–Cl	800–700	Difficult to interpret
C–Br, C–I	Below 650	Often out of range of instrumentation

(continued)

Table 2 (continued)

Bond	Wavenumber/ cm^{-1}	Notes
Bending vibrations		
R–N–H	1650–1500	Take care not to confuse N–H bend with the C=O stretch in amides
R–C–H	1480–1350	Saturated alkanes and alkyl groups
R–C–H	1000–680	Unsaturated alkenes and aromatics

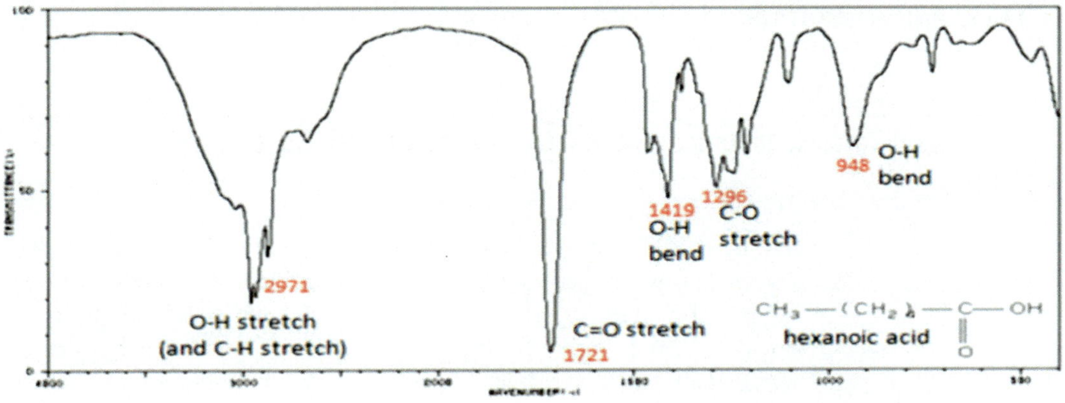

Fig. 13 An infrared spectra

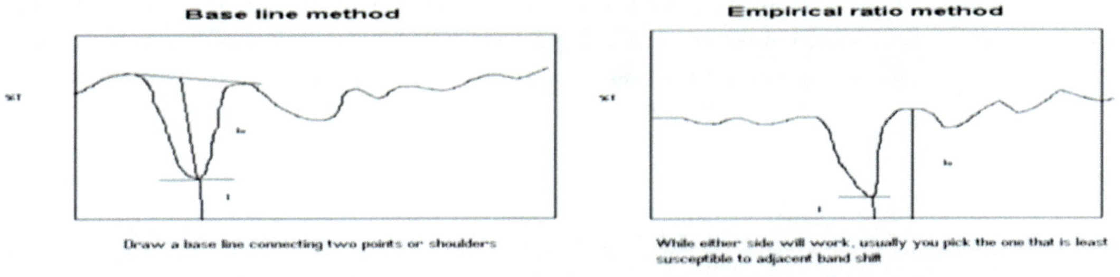

Fig. 14 Plot the log of I_0/I vs. concentration to establish a calibration curve

7.3 Placement of IR Analyzer in a Dairy Industry

IR analyzer is used for the process monitoring and control and can be placed in different modes as follows:

- Off-line: Away from the process stream/equipment—quality assurance lab.
- At-line: Nearby process—sample removed—lab on raw milk reception dock.
- On-line: Side stream of the main process stream—loop in side stream.
- In-line: Interface to process stream—main stream.

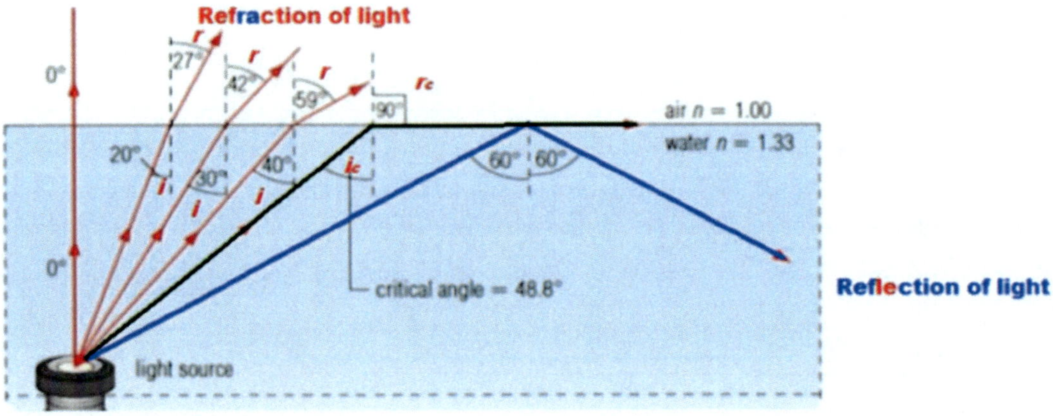

Fig. 15 Total internal reflection [9]

8 Attenuated Total Reflectance-Fourier Transform Infrared Spectroscopy (ATR-FTIR)

As the radiation moves form the high RI medium (water) to low RI medium (air), it gets refracted. With increase in the angle of incidence, the angle of refraction also increases. At a certain angle of incidence, the angle of refraction becomes 90° which is known as critical angle of incidence (Fig. 15). When angle of incidence (i) becomes greater than the critical angle of incidence (ic), the incident radiation is completely reflected back in the same medium which is known as total internal reflection.

Attenuated total reflectance spectroscopy is the most versatile and powerful technique for IR spectroscopy. A large number of solid and liquid samples directly examined without preparations. In modern IR spectrometers, ATR accessory mounting is in the sample compartment. Sample accessory (internal reflection element) of ATR FT-IR spectroscopy is the crystal (prism) having RI significantly greater than the sample.

8.1 Basic Principle of ATR Spectroscopy

Basic phenomenon is the total internal reflection in which IR radiation is focused on one end of ATR crystal at an angle > critical angle of incidence and the radiation is reflected back into the crystal. At interface, IR radiation penetrates into the sample up to short distance called as penetration depth (dp) which typically ranges from 0.5 to 3 μm. The IR radiation penetrated into the sample is called as evanescent wave (Fig. 16). The evanescent wave interacts with analyte molecules and loses its intensity (attenuated) and returns back to the ATR crystal and exits from other end of the crystal. It then reaches IR detector and generates IR spectrum. Changes which occur in IR radiation on interacting with analyte molecules are measured from the IR spectrum.

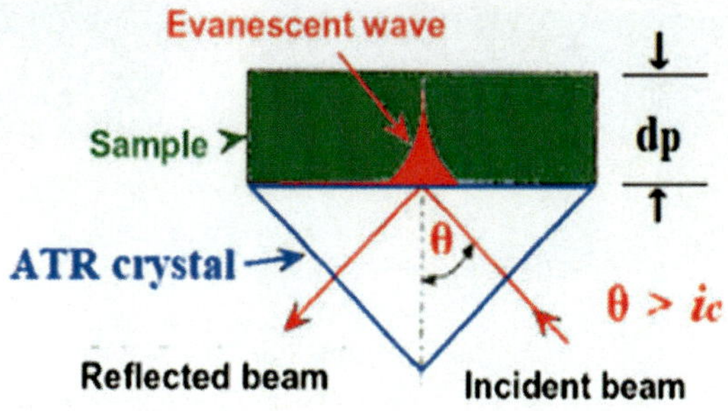

Fig. 16 Evanescent wave [9]

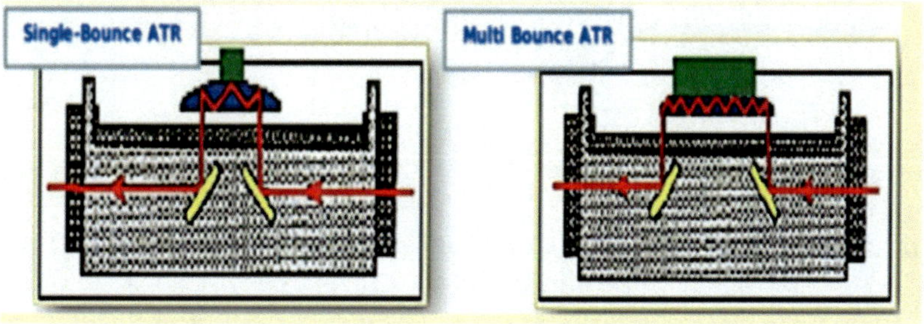

Fig. 17 Single-bounce and multibounce ATR [9]

8.1.1 Configurations of ATR Accessory

ATR accessories are available with two configurations: single-bounce and multiple bounce (Fig. 17).

Effective path length or the sampling area will be determined by the number of reflections and the penetration depth of evanescent wave. Number of reflections again will depend upon geometry of crystal—thickness, length, as well as the angle of incidence.

8.1.2 Properties of Single-Bounce ATR (SB–ATR) (Fig. 17)

- Only single internal reflection.
- Limited contact with sample.
- Narrow sampling area.
- Produce weak spectra.
- Suitable for strong absorbers and concentrated solutions.

8.1.3 Multiple Bounce ATR (MB-ATR) (Fig. 17)

- Multiple internal reflections—up to 25, but 9 or 11 most common.
- Greater contact with sample.
- Broad sampling area.
- Produce intense spectra.
- Useful for weak absorbers and dilute solutions.

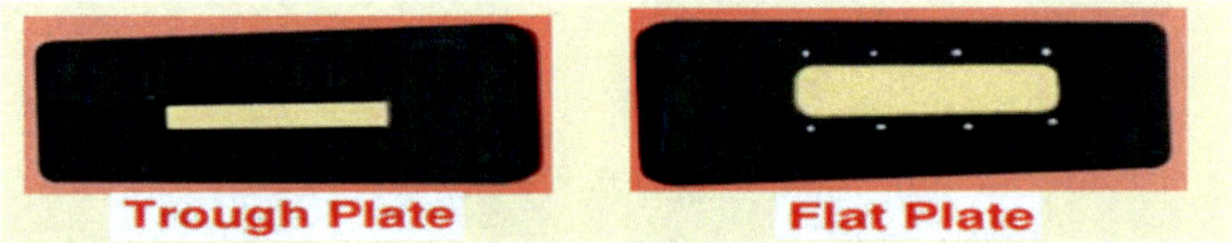

Fig. 18 Types of horizontal ATR [9]

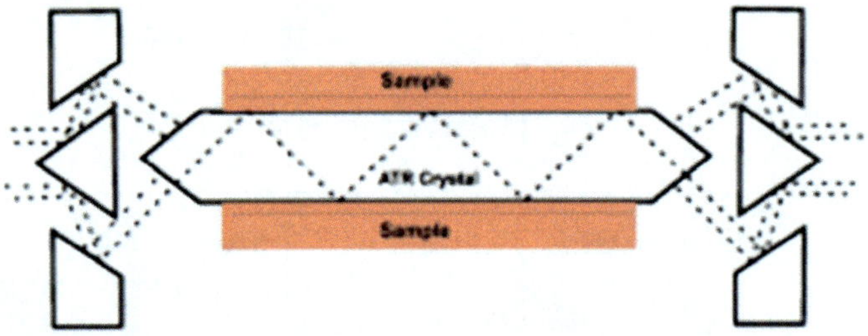

Fig. 19 Cylindrical ATR [9]

8.2 Designs of ATR

ATR cells have different designs like traditional vertical face, horizontal, and cylindrical.

8.2.1 Traditional Vertical Face ATR

- In traditional vertical face ATR a thin sample is clamped against vertical crystal and the face is replaced by more modern designs.

8.2.2 Horizontal ATR (HATR) (Fig. 18)

In horizontal ATR, crystal plate or side is of 5 cm × 1 cm with the upper surface exposed.

Types of samples which can be measured by horizontal type ATR.

- Solid sample.
- Cloth, film, yarns, and paper.
- Coating films on metals or resins.
- Gel and liquid samples.

8.2.3 Cylindrical ATR (Fig. 19)

- Limited to mobile fluids.

8.3 ATR Crystals Materials

Selection of the ATR crystals materials depends upon the wavenumber transmission range, chemical properties of the sample and crystal, physical properties of materials, and cost of materials. Table 3 shows the characteristics of different materials used in ATR crystals. The most commonly used materials in ATR crystal are:

8.3.1 Diamond

Wavenumber range: 40,000–12.5 cm^{-1}.

Refractive index: 2.38.

Table 3
Characteristics of different ATR crystals

Type of material	Wavenumber range (cm^{-1})	Refractive index	Solubility in water	Features
KRS-5	20,000 ~ 290	2.37	Practically insoluble	Orange, soft, poisonous
Ge	5000 ~ 700	4.0	Insoluble	Silver, fragile
ZnSe	20,000 ~ 650	2.4	Insoluble	Yellow, hard
Diamond	40,000 ~ 400	2.38	Insoluble	Transparency, hardest

Water solubility: insoluble.
Hardness: very hard, 10,000 kg/mm^2.
Diamond ATR crystal is being preferred as:

- Better contact of the sample with the diamond ATR due to its better mechanical strength which allows the compression of the sample and also reduces the amount of sample required for analysis.
- It is resistant to all corrosive and rough solvents and samples.
- Lower coefficient of friction of diamond makes it easy to clean.

8.3.2 Zinc Selenide (ZnSe)-Easy to Scratch

- Wavenumber range: 10,000–550 cm^{-1}.
- Refractive index: 2.40.
- Water solubility: insoluble.
- Hardness: 250 kg/mm^2.

8.3.3 Germanium (Ge)-Fragile

- Wavenumber range: 10,000–550 cm^{-1}.
- Refractive index: 2.40.
- Water solubility: insoluble.
- Hardness: 780 kg/mm^2.

8.4 Advantages of ATR IR Spectroscopy

Revolutionized analysis of solid and liquid samples.

- Combat most challenging aspects of IR analyses.
- Quick or no sample preparation.
- Better reproducibility of the attained spectra.
- Excellent sample-to-sample reproducibility.
- Better accuracy with reduction in variations occurring due to operator.
- Non-destructive analysis.

8.5 Application of ATR IR Spectroscopy

ATR can be used to analyze:

- Solid, semisolid, and liquid states.
- Gel, paste, sheet/film, powder, etc.
- Nonaqueous solutions—oils, solvents, inks etc.

Excellent technique—measuring composition of solids.

- Powder, granule, tablet, etc.
- Pure or mixtures.
- With little or no sample preparation—nondestructive.
- Liquid sample.
- Pouring small amount over surface of crystal—sufficient. ATR-FTIR was used as an alternative technique to assess proteolysis in ultra-high temperature (UHT) milk [2].

9 Application of Chemometrics to Develop Spectroscopic Method

Chemometrics differs from the classic statistics by two ways:

- Takes multiple variables concurrently.
- Considers collinearity of the variables: variation in one variable or group of variables with respect to the covariation of other variables.

Application involves

1. Training set.
2. Acquisition, preprocessing. and selection of variables.
3. Calibration of model (development of prediction model).
4. Validation of prediction model.

Entire procedure in application of chemometric to develop method is divided into series of straightforward logical steps.

1. Selection of a training set.
   (a) Qualitatively and quantitatively known references samples, for which variables to be measured.
   (b) For application to detect adulteration in Ghee:-Samples of pure and adulterated ghee and foreign oils and fats.
2. Acquisition of variables.
   (a) Contain qualitative and quantitative information of the references samples.
   (b) For application to detect adulteration in Ghee: FT-NIR spectra of pure and adulterated ghee and foreign oils and fats.

3. Preprocessing of the variables.
    (a) Reduced random noises and enhanced spectral features—reduction or omission of unwanted sources of variation. It can be done by baseline correction, mean centering, scaling, and absorbance of normalized FT-NIR spectra.

9.1 Selection of Variables

Those variables which contain information of the references sample for aimed classification are retained, while those deprived of any discriminating power or as a noise are removed.

For application to detect adulteration in Ghee: Spectral features of ghee and foreign oils and fats.

Use of suitable software—designed for automatic selection of variables.

9.2 Calibration of Model

Building a mathematical relationship between selected variables taken as training set and while their known categories or quantitative values of sample property.

Relationship between spectral features of pure ghee and those of foreign oils and fats at varying degree of adulteration.

- The suitable software designed from streamline process for automatic models building and determining the acceptance thresholds.

9.3 Review of Classification Results

Indicates issues with data or performance of the method.

- Flagged by troubleshooting engine.
- Allows corrective action to be taken, if any.

9.4 Validation of the Model

Evaluate reliability of classification achieved using

- Cross validation.
- External validation.

9.5 Selection of Algorithm

The most commonly used in multivariate calibration.

- Partial least square (PLS) and principal component regression (PCR).

9.6 Validation of Calibrated (Prediction) Model

Both approaches of validation.

- Cross validation

    Samples of training set used.
- External validation

    Independent test set of samples prepared.

References

1. Nielsen SS (2017) Introduction to food analysis. In: Food analysis. Springer, Boston, MA, pp 3–16
2. Ranvir S, Sharma R, Gandhi K, Upadhyay N, Mann B (2020) Assessment of proteolysis in ultra-high temperature milk using attenuated total reflectance–Fourier transform infrared spectroscopy. Int J Dairy Technol 73(2): 366–375
3. https://www.shimadzu.com/an/service-support/technical-support/analysis-basics/ftirtalk/talk9.html
4. https://www.ifsc.usp.br/~lavfis2/BancoApostilasImagens/ApLuminescencia/Infrared%20Spectroscop1.pdf
5. https://chemistryscore.com/how-to-interpret-ir-spectra/
6. https://personal.utdallas.edu/~scortes/ochem/OChem_Lab1/recit_notes/ir_presentation.pdf
7. https://thefactfactor.com/facts/pure_science/physics/total-internal-reflection/6985/
8. https://old.vscht.cz/anl/vibspec/FTIR%20Reflection%20Techniques.pdf
9. https://www.slideshare.net/suraj_mindgamer/micro-atr

Check for updates

Chapter 10

Mass Spectroscopy

Abstract

Mass spectroscopy is an analytical technique used to measure *m/z* ratio of the charged particles (ions). Mass spectrometry can be used to determine the exact mass of the molecules ranging from large molecules like proteins to small molecules in the natural substances. In this chapter, the basic principle and steps involved in mass spectroscopy have been covered. The principle and working of ionization methods like matrix assisted laser desorption ionization (MALDI), electrospray ionization, atmospheric pressure chemical ionization, atmospheric pressure photoionization and mass analyzers like magnetic sector, time of flight (TOF) mass analyzer, triple quadrupole mass analyzer, and ion trap mass analyzer have been explained in detail. Various parameters used in acquisition like ion spray voltage, declustering potential (DP), curtain gas, CAD (collisionally activated dissociation) gas, collision energy (CE), collision cell exit potential, dwell time (DT), and cycle time have been discussed. The concept of tandem MS has also been touched upon. Advantages of mass spectrophotometer over other detectors have also been emphasized.

Keywords Matrix assisted laser desorption ionization (MALDI), Electrospray ionization, Atmospheric pressure chemical ionization (APCI), Atmospheric pressure photoionization (APPI), Mass analyzers like magnetic sector, Time of flight (TOF) mass analyzer, Triple quadrupole mass analyzer, Ion trap mass analyzer, Ion spray voltage, Declustering potential (DP), Curtain gas, CAD (collisionally activated dissociation) gas, Collision energy (CE), Collision cell exit potential, Dwell time (DT), Cycle time

Mass spectroscopy is an analytical technique used to measure *m/z* ratio of the charged particles (ions). Mass spectrometry can be used to determine the exact mass of the molecules ranging from large molecules like proteins to small molecules in the natural substances. In mass spectrometry, mass of a particular molecule is determined, but it is not possible to know what particular protein is it. Thus, mass spectrophotometer is a sophisticated weighing machine. So, we have to digest the protein into smaller peptides and those peptides will be subsequently introduced in mass spectrometer which can be identified by using various software. Apart from this, small molecules can also be extracted from the biological sample which can be used for its identification and quantitation. A biological mixture of samples may contain hundreds of proteins. In clinical proteomics aspects, first the protein is identified and then

Kamal Gandhi et al., *Advanced Analytical Techniques in Dairy Chemistry*, Springer Protocols Handbooks,
https://doi.org/10.1007/978-1-0716-1940-7_10,

the biomarkers are determined. Identifying various biomarkers for various physiological and bilateral processes has been a recent trend where individual biomolecules can be identified and subsequently can be used for diagnosis of various diseases/disorders. As such one can identify complex of proteins which can be further analyzed by using various bioinformatics software.

1 Principle of Mass Spectroscopy

Mass spectrometry is the study of multiple molecules based on their masses. A spectrum of distinct masses may be collected and compared to existing masses or a pattern of masses obtained from the theoretical breakdown of the whole compound. Mass spectrometry may be used to identify and classify a wide variety of substances. The first work on mass analysis date all the way back to the early decades of the twentieth century when JJ Thompson won the 1906 Nobel Prize for inventing a prototype for measuring the mass-to-charge (m/z) ratio. The idea that m by z can be studied was applied for the first time in experimental experiments to determine the mass of elements by Francis Aston, a Cambridge physicist, in 1919, for which she was awarded the Nobel Prize in 1922. The mass-to-charge ratio of ions is determined in mass spectrometry in order to classify and quantify molecules found in different samples.

2 Steps Involved in Mass Spectroscopy

1. Introduction of sample molecules to the mass analyzer by HPLC.
2. Ionization of the analyte molecules to the charged particles.
3. Separation of these ions according to m/z ratio of these ions.

The analyte molecule is rendered negatively charged by ionization. Heavier mass which is charged would therefore be deflected less under an electric field and thus move or spin more slowly. The isolated ion mass may be transformed to various electronic signals that can be detected using an appropriate detector. Different masses can pass through the mass spectrometer and their relative frequency can be determined by the height of the peak, which is the amount of specific signals obtained at a given time point.

Mass spectrometers consist of three parts: (a) an ion source, (b) mass analyzer, (c) detector system. The sample is kept and ionized in the ionizer which then enters into the mass analyzer and the mass of individual compounds is analyzed which traverses through the mass analyzer. The ions are then hit the detector and gives the signal in the form of a spectrum which is captured by the computer and deconvoluted in order to identify these molecules.

3 Methods of Ionization

There are many methods of ionization, the two main classifications are fast atom bombardment which is of semi-hard type and the second one is electron impact that is a hard method. For proteomics purposes, fast atomic bombardment (FAB) is most suitable which includes MALDI (matrix assisted laser desorption ionization), ESI (electrospray ionization), APPI (atmospheric pressure chemical ionization), and APPI (atmospheric pressure photon ionization). Ion source can also be categorized into three categories:

1. Those requiring sample in the gas phase prior to the ionization: electron ionization and chemical ionization.
2. Those for low volatile and high molecular weight samples: operating with sample solution: ESI, APCI, and APPI.
3. Those based on sample desorption from a solid substrate: MALDI, LDI.

3.1 Matrix Assisted Laser Desorption Ionization (MALDI)

In matrix assisted laser desorption ionization (MALDI), the protein is first digested into peptides which are then mixed with the matrix and allowed to crystallize on a plate (ratio of the sample to the matrix to be kept at 1:10000 to prevent the ionization of the analyte by the laser). The laser is fired at the matrix which produces the ions (Fig. 1). Some of the matrices used are 3–5 dimethoxy-4-hydroxycinnamic acid (sinapinic acid), α-cyano 4-hydroxy cinnamic acid, 2–5 hydroxy benzoic acid. When the laser gets activated, the energy released from here heats the MALDI plate which is absorbed by the matrix which in turn gets transferred to the analyte. As a result, the analyte is ionized into gas phase and the gas phase analyte travels in the electric field. In the case of MALDI, the analyte molecules are co-crystalized with a matrix, which absorbs in the wavelength of the laser, and this is spotted on a metal plate (target). The laser pulse results in rapid heating of the sample, leading to a microexplosion that ejects a plume of matrix and sample away from the target. NALDI (nanostructure-assisted laser desorption/ionization) is somewhat similar to MALDI. For both methods, ionization occurs by illumination of a solid sample

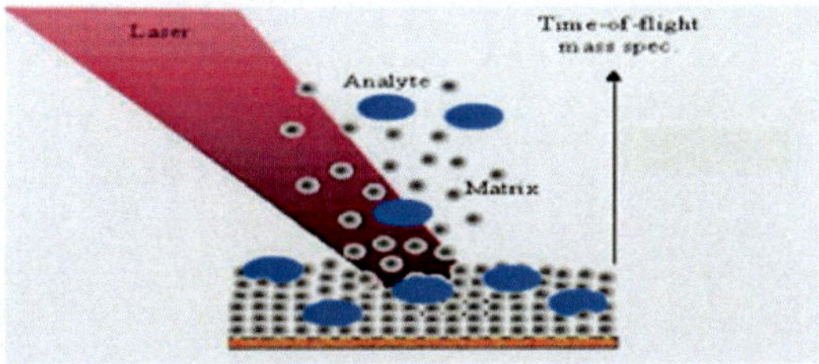

Fig. 1 Principle of MALDI [1]

with a short laser pulse. In NALDI, ionization is believed to occur just above the active layer of the NALDI surface by proton transfer reactions between analytes and protonated species originating from the active layer that are generated through a series of photochemical reactions [10].

Steps in MALDI:

1. First matrix analyte solution is prepared.
2. Mixing together well homogenized solution to have the cocrystals.
3. Excitation of the cocrystal by laser. Matrix will be excited and transfer this energy to the analyte. Ions will be transferred to the mass analyzer.

Characteristics of the ideal matrix:

1. It should absorb a large amount of laser energy and ionize the analytes indirectly.
2. Serve as a proton donor or receptor to ionize analytes in positive/negative modes.
3. In the presence of the dedicated matrix, samples easily ionize: If insufficiently suitable matrices or even combinations of the matrices are used, the analysis would fail.
4. Using the right matrix will result in a strong signal-to-noise ratio, a well-resolved spectrum, and reduced analyte fragmentation.
5. It can have a higher laser absorption than the laser's emission wavelength, which is usually in the UV range of 337 nm or 355 nm. The nitrogen laser emits a wavelength of 337 nm, which may later be replaced by Nd YAG (neodymium-doped yttrium aluminum garnet laser) since the former has a shorter lifespan and higher photon intensity (337 nm), which may weaken the matrix and may not be absorbed by the matrix. All known organic matrices have aromatic ring system with π electrons that are delocalized that will increase the absorption coefficient of the matrix (for laser absorption) and so ionization will improve.
6. Matrix should be low in mass to be easily vaporized upon laser incidence and must be low volatile or it will evaporate spontaneously.
7. Matrix should be polar to be easily soluble in polar solvents (water, methanol, acetonitrile etc.) and then it will be mixed with the samples easily to create an aqueous matrix analyte solution.

8. Matrix should be acidic in nature so that it can act as a proton source for the ionization of the analytes, for example, benzoic acid or cinnamic acid.
9. The matrix should remain as stable as practicable under high vacuum conditions, as the MALDI ionization phase happens under high vacuum conditions ($1 \times 10^{-8}/1 \times 10^{-9}$ bar).

MALDI can be used in both the positive or the negative mode. In positive mode, hydrogen ion or the sodium ions will be exchanged between the matrix and the analytes to form the adducts or quasi-molecular ion. In case of the negative mode, hydrogen ion or the sodium ions are extracted from the analytes to form negative ions. As the negative mode is more selective, so positive mode is more popular.

Choice of the matrix is the key to obtain a good spectrum:

There are three different classes of matrices:

1. Classic organic matrix: Benzoic acid, cinnamic acid, or derivatives.
2. Liquid crystalline matrices.
3. Inorganic matrices such as graphite.

Dried Droplet Method

1. Prepare a fresh saturated solution of the matrix (10–20 mg matrix powder +1 ml of solvent in a 15 ml of Eppendorf tube) and centrifuge to dissolve the undissolved matrix.
2. Place the supernatant in another tube.
3. Add a smaller volume of sample solution (1–2 ml) to the matrix.
4. Mix thoroughly for a few seconds in a vortex mixture.
5. Place a 0.5–2 μl of droplet mix on MS sample stage.
6. Allow the droplet to dry at the room temperature. When the droplet will be completely evaporated and dried, the sample is loaded in MALDI plate.

Dried droplets are very stable and can be kept under vacuum or dark place for days before running.

3.2 Electrospray Ionization

The second ionized protocol is electrospray ionization, the sample in the form of peptides which have been generated or which have been purified is mixed with the volatile solvents (acetonitrile and methanol) which are allowed to enter into the source and are forced to be charged droplets which come out of the tip of the needle and subsequently evaporated leaving the charged samples. Electron ionization is not selective and there occurs extensive fragmentation in it. First there occurs production of charged droplets at the electron spray capillary tip: Nebulizer or sheath gas (N_2) shears

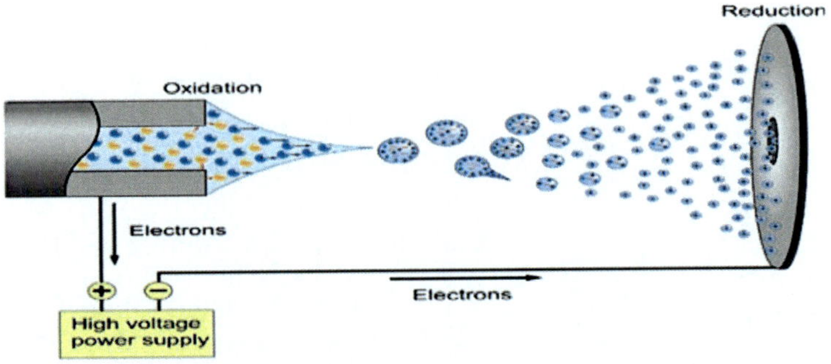

Fig. 2 Electron spray ionization (Source: Wikipedia [2])

around the eluted sample (elute carrying the sample molecules) and by the aid of the electric field applied (2–6 eV), highly charged droplets with the same polarity as the voltage applied are produced at the exit of the electrospray tip. Charged droplets will be evaporated by temperature effect and drying gas to remove the solvent molecules. There occurs decrease in the size of the droplets and charge density. Repulsion between the like charges will be higher as compared to the surface tension forces of the droplets. This will result in the breaking of the droplets to smaller and smaller droplets resulting in the ejection of the ions which are transported to the mass analyzer. Electron spray ionization (Fig. 2) can be used in both positive and negative modes. In positive mode, we will have the positive ions, while in the negative mode, negative ions will be generated.

3.3 Atmospheric Pressure Chemical Ionization

Following are the steps involved in the atmospheric pressure chemical ionization for the production of the ions:

1. Eluent carrying sample molecules will pass through the spray needle. Nebulizer gas or sheath gas (N_2) shears around the molecules to spray droplets (Fig. 3).
2. Desolvation process: Sample molecules will then pass through a heated vaporizer (200–400 °C) and by the aid of this heater and desolvation gas (N_2), solvent molecules will be evaporated resulting in the analyte molecules and solvent vapors.
3. Corona discharge produces electrons that will ionize the solvent molecules to form solvent ions. Solvent ions react with other solvent molecules to form ions.
4. Chemical ionization: Solvent ions produced SH/S^- will transfer charges to the analyte molecules to form protonated species or deprotonated species. In positive mode, proton will be transferred to the analyte molecules resulting in positive ions, while in the negative mode, proton loss or electron transfer will take place resulting in the negative ions.

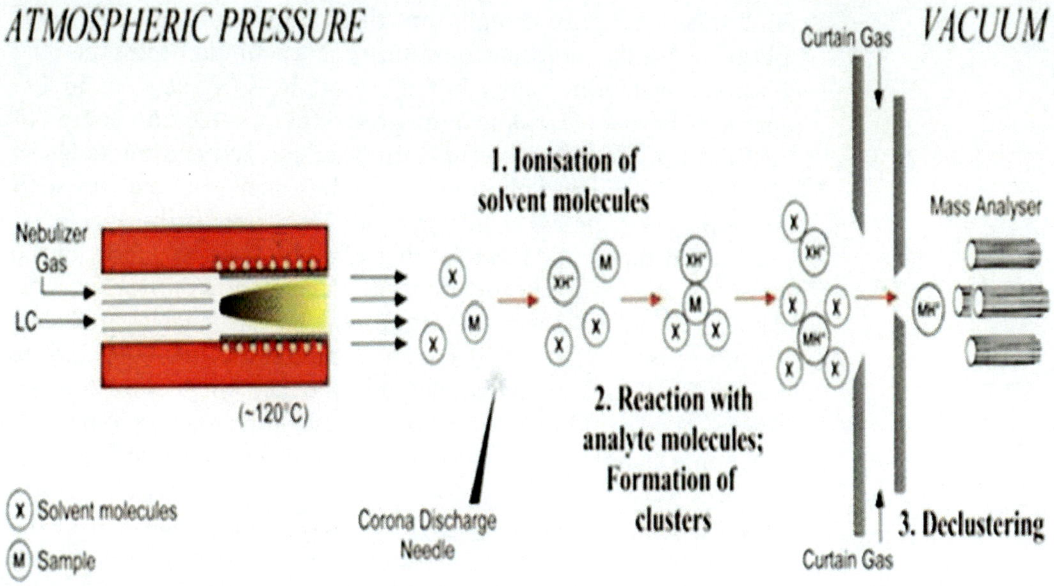

Fig. 3 Atmospheric pressure chemical ionization [3]

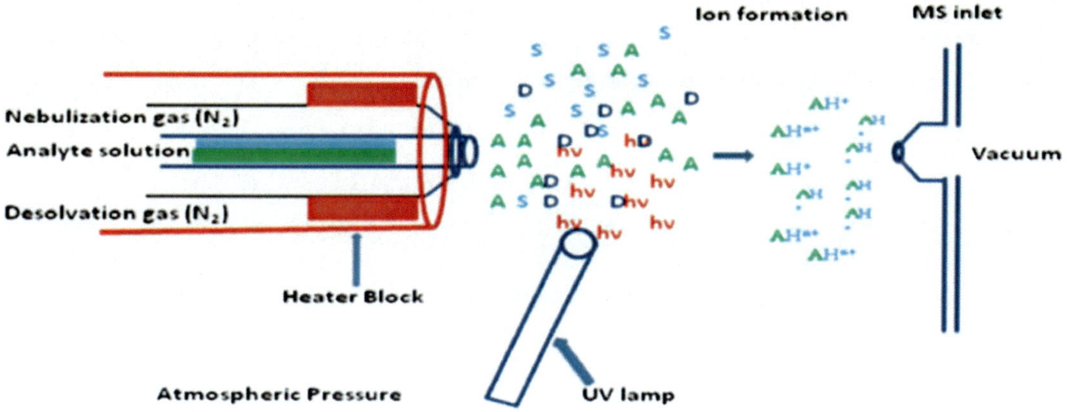

Fig. 4 Atmospheric pressure photoionization [4]

3.4 Atmospheric Pressure Photoionization

Steps Involved in the APPI

- Eluent carrying sample molecules will pass through heated capillary or spray needle.
- Nebulizer gas (N_2) will shear around the sample molecules converting around them to spray droplets.
- Desolvation by discharge gas and temperature (heater) to evaporate the solvent molecules which will result in the generation of analyte molecules and solvent vapors.
- Photons generated with UV lamp, electric discharge lamp with energy up to 10 ev which is sufficient to ionize the analyte molecules (Fig. 4).

If the ionization energy of the molecules is higher than the ionization energy of the photon, these cannot be ionized by the photons. So, the molecule ionization energy should be less than the photon ionization energy but most of the solvents used in LC methods have ionization energy greater than ionization energy of the photon. So, these solvents cannot be ionized and the range of the energy of these photons can only ionize the analytes with ionization energy less than that of the photons. If the analyte M has ionization energy less than that of the ionization energy of the photon, cation will be generated with the emission of the electrons. This is called as direct ionization. But if the analyte molecule absorbs photon hv and becomes excited M*, then the analyte molecule releases energetic electron and becomes radical cation. This is called as indirect ionization. In case of the indirect ionization wherein it is not possible to ionize the molecules, we make use of the dopants which will promote the ionization of the analyte molecules. Dopants will have ionization energy lower than that of the photons. So, it will be easily ionized by the photons. Dopants will produce the analyte ion by proton transfer reaction.

$$D + h\nu \rightarrow D^+$$

$$D^+ + A \rightarrow A^+ + D$$

Another way is that the solvent ion will be ionized by the dopant to form the solvent ions.

$$D + h\nu \rightarrow D^+$$

$$D^+ + S \rightarrow [D - H] + [S + H]^+$$

$$[S + H]^+ + A \rightarrow S + [A + H]^+$$

Analyte ion will then be transferred to the mass analyzer.

4 Mass Analyzers

Now, the ionized substances have been produced but once these substances or samples have been ionized, they have to be entered into the mass analyzer where the mass of these samples will be analyzed. This is the segment of the instrument which separates the ions based on its *m/z* value that is mass by charge ratio. Mass analyzer is used to get the structural information of the compound by fragmentation. All molecular ions formed in the ion source and transferred to mass analyzer will be fragmented to form a specific product ion. Mass analyzer improves the selectivity because it can differentiate between two or more compounds having the same spectra. Fragmentation: each molecule has a specific product ion. Mass analyzer can also improve the sensitivity because it allows only

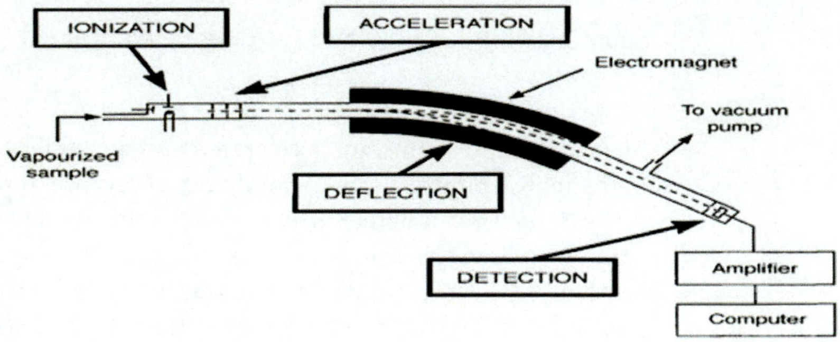

Fig. 5 Magnetic sector

selected m/z ions to pass through the detector but others which do not have a stable trajectory cannot pass to the detector.

4.1 Types of Mass Analyzer

Some of the common types of mass analyzers include: time of flight, quadruple analyzer (low resolution but it is fast and cheap), ion trap mass analyzer (good resolution, all in one mass analyzer), magnetic sector analyzer (MSA) (high resolution and determines the exact mass), ion cyclotron resonance (highest resolution, determines exact mass but it is costly). These different types of analyzers can be put in tandem in various combinations and modification, thus getting different nomenclatures like triple quadruple, MALDI-QqTOF, QqTOF, ESI-OTOF, LC-ESI-MS/MS, etc.

4.1.1 Magnetic Sector [5]

The magnetic sector contains only a magnetic field which separates ions according to their m/z ratio (Fig. 5); however, the resolution will be limited by the fact that ions leaving the ion source do not always have the same kinetic energy and therefore do not all have the same velocity. Therefore, to improve resolution, ions leaving the ion source will be accelerated to a high velocity and their kinetic energy will be measured.

$$\mathrm{KE} = 1/2\ mv^2 = eV$$

where m denotes the ion's density, v denotes the ion's velocity, and e denotes the ion's electric charge. V stands for accelerated voltage.

There are two forces acting on the ions: the Lorentz force equals BeV and the centrifugal force equals mv^2/r. These two forces must be balanced in order to maintain the ions' path across the magnetic field.

$$F_1 = BeV$$

$$F_2 = mv^2/r$$

To move through the magnetic field area and enter the detector, these ions must take a long, curved path with a defined radius (r), where

$$F_1 = F_2 \qquad BeV = mv^2/r$$

Squaring on both sides and rearranging we will get

$$m/e = B^2r^2/2V$$

By maintaining a constant accelerated voltage while varying the magnetic field strength, or by maintaining a constant B while varying V, the detector can detect provided m/z ions. Ions with varying m/z values follow the same direction across the magnetic field before they reach the detector. Each scan would result in the generation of a single mass spectrum. Ions with a lower mass would be deflected first, followed by ions with a higher mass. Thus, the magnetic sector analyzer can discriminate between ions depending on their m/z ratio, but precision is restricted due to the fact that ions exiting the ion source do not always have the same kinetic energy and thus travel at varying speeds. As a result, identical ions will not arrive at the detector simultaneously. As a consequence, sensitivity and resolution can suffer. In general, as related ions move through a magnetic field, they are all deflected to the same degree and all go in the same direction, but their kinetic energy is not identical. As a result, they would not arrive at the detector simultaneously. By focusing the ions according to their kinetic energy, an electric field is used to improve resolution. When an electric field is applied, ions of the same kinetic energy arrive at the detector simultaneously. As a result of going through the electrostatic analyzer's electric field, which is applied perpendicular to the ion motion, the ions are deflected in a radial direction. Ions with the same mass-to-charge ratio travel at the same velocity and arrive at the detector simultaneously, resulting in high precision for each m/z ratio.

4.1.2 Time of Flight (TOF) Mass Analyzer

Linear TOF comprises acceleration chamber, flight tube with known length, and detector.

Principle of Operation

Separation of ions is based on the time it takes for ions to travel through the flight tube and reach the detector. Acceleration voltage will be applied to all the ions at the same time to accelerate them through the flight tube (with known length) to the detector. So, velocity of these ions depends only on their m/z ratio where lighter m/z ions travel faster to detector than heavier ions and the time it takes for the ions to travel through flight tube determines the m/z ratio of these ions.

Kinetic energy for the ions (T) is given by

$$T = ZV = \frac{mv^2}{2} \tag{1}$$

where

Z = charge of ions.

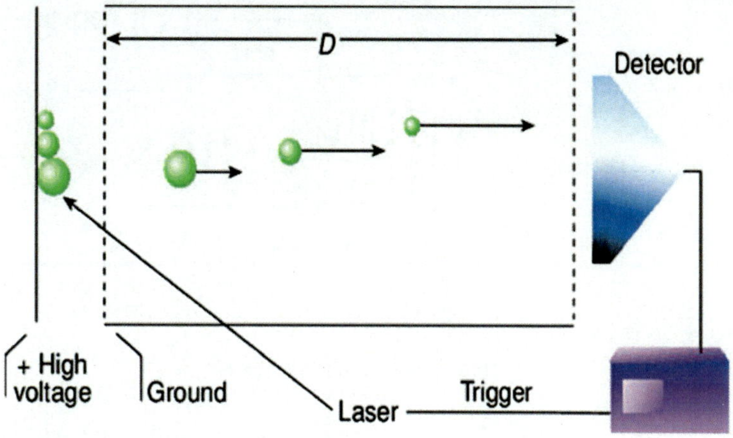

Fig. 6 Time of flight mass analyzer [6]

v = velocity of ions.
M = mass of the ions.
V = acceleration voltage.
Putting the value of $v = L/t$ in Eq. 1, we will get:

$$T = \frac{1}{2} mL^2/t^2$$

$$ZV = 1/2mL^2/t^2$$

Acceleration voltage is constant for all ions and also L is constant

Therefore, $m/z = kt^2$

Thus, separation time in flight mass analyzer depends upon the time it takes for the ions to travel in tube. Higher is the m/z ratio of the ions, more time it will take for separation in the tube. Lighter ions will travel faster to the detector than the heavier ions (Fig. 6). Resolution is poor in linear TOF mass analyzer because starting time and kinetic energies for all the ions are different. So, ions leaving the ion source have different kinetic energies, different starting location, and time. So, velocity of ions and arrival time will be different. So, ions with the same m/z ratio also have different arrival times and that will cause poor mass resolution and low sensitivity.

To compensate for these differences in TOF, reflectron (Fig. 7) is added at the end of the flight tube which acts as a focusing mirror to reflect ions in the opposite direction to the detector. So, by applying reflectron in TOF, ions having the same m/z but different kinetic energy and start time will reach the detector at the same time because higher energetic ions will penetrate more deeply in the reflectron and take longer time before turning again to detector and they will reach the same time with lower energetic ions of same m/z ratio and will improve the resolution and increase sensitivity.

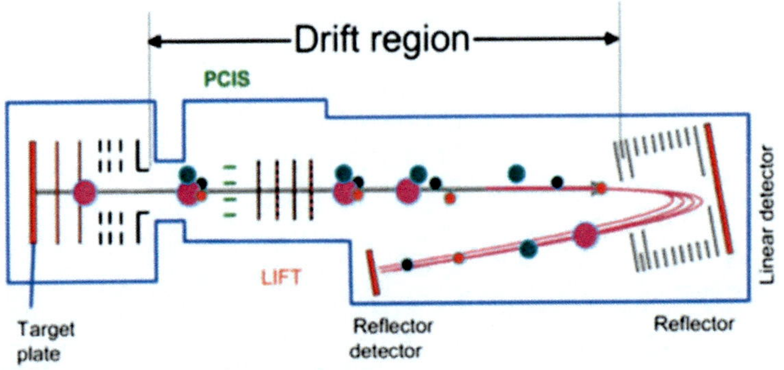

Fig. 7 Reflectron TOF [6]

Recent instruments are configured with a second mirror. So, ions transverse the flight draft region four times before detection (W_{mirror}) and decrease the total drift length and hence the resolution increases. This also results in the use of the shorter flight tube design as compared to longer design used earlier.

In TOF, molecules are classified according to their molecular weight. On the basis of the *m/z* ratio, the ions liberated from the ionizer and accelerated through the vacuum of the mass analyzer are isolated. Different sized molecules need varying amounts of time to enter the detector. The whole machine resides in a state of extreme vacuum. Throughout the protocol, the signals that strike the detector are intensified, and the ions are then analyzed. Thus, the relative concentration of ions is determined by their mass-to-charge ratio and associated intensity. To the extent that possible, the device has a high-throughput and no upper limit on its resolution. Thus, TOF has been adapted to the application at hand; for example, TOF in triple quadrupole is well suited for amino acid sequencing, while MALDI TOF is well suited for both amino acid sequencing and molecular weight determination. QTOF is an effective method for sequencing amino acids and modifying proteins.

The quadruple mass filter (Fig. 8) is composed of four parallel metal rods with polarized charges: two opposite rods receive a positive potential, while the other two rods receive a negative potential. Voltages applied to the ion's flight path have an impact on the ion's trajectory. At a given voltage, some entities entered the detector, while others met with the rods and were unable to enter. As a consequence, by scanning the sample through a large voltage range, different *m/z* entities may be selected. It is distinguished from time of flight in that voltage is used to distinguish ions, while in TOF, time of flight is used to separate ions. The quadrupole mass analyzer is composed of four parallel rods organized in a square configuration, two of which are at positive potential and two of which are at negative potential. DC and RF or AC are applied in

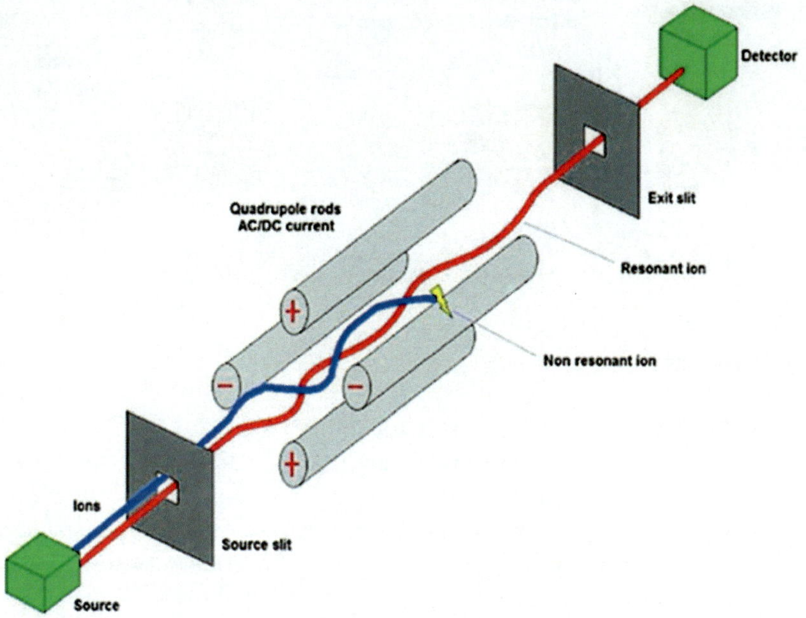

Fig. 8 Quadruple mass analyzer [7]

combination to these rods, creating an electromagnetic field that determines the *m/z* ratios travel through the filter to the detector. Thus, by combining DC and RF, only ions with a particular *m/z* ratio are allowed to travel through the filter to the detector, avoiding collisions with the rods (resonant ions) and interfering ions that do not have a stable trajectory to the detector will hit the rods and cannot pass to the detector (non-resonating ions) (Fig. 7). If the ions charge is large there is a strong force on ions to be able to transfer large ions to the detector. So, magnitude of DC and AC increases with time and magnitude of AC is six times strength of DC but the ratio of RF/DC is constant because each pair of rods are connected. Rods have exactly the same voltage as the one directly opposite. If RF > DC, smaller *m/z* ions will start bombarding to the wall and distance to the detector will increase but the larger ions will bombard less and move faster. So, larger *m/z* ions that have a higher mass will reach the detector first and then the smaller ions. But if the DC > AC, smaller *m/z* ions will migrate faster to the detector and by applying both of them on the rods, electromagnetic field will be generated that will determine which *m/z* ratio will pass first to detector based on the magnitude of DC and RF. By applying DC and RF voltage between one pair of rods and the other will cause the movement of the ions. As they move, positive ions will migrate to the positive electrode and they will never mix up.

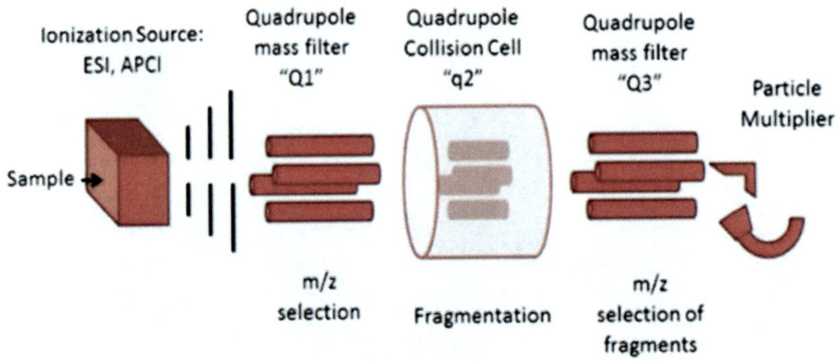

Fig. 9 Triple quadrupole mass analyzer [8]

4.1.3 Triple Quadrupole Mass Analyzer

It consists of three quadrupoles Q1, Q2, Q3. Q1 is the mass filter, Q2 is the collision cell, and Q3 is again a mass filter (Fig. 9).

Q1 mass filter acts as a separation device to select a specific precursor ion (targeted ions). So, only selected *m/z* ions which have a stable trajectory can pass through Q1 to Q2 and so spectrum will not be complicated and sensitivity will increase.

Q2 (collision cell or CID (collision induced dissociation)) for fragmentation of ions by accelerating of selected precursor ions in the presence of collision gas by the aid of temperature and voltage applied (collision energy). Collision energy can be varied to allow different degrees of fragmentation. Each precursor ions will be converted to the product ions. For each precursor ions we will select two product ions, one of which will be used as qualitative and another one as quantitative.

Q3 The product ions will be transferred to the Q3 (mass filter). Q3 acts as a separation device and also for filtering only specific product ions for each parameter.

4.1.4 Ion Trap Mass Analyzer

The third most frequently used ion trap mass analyzer traps and analyses ions using a three-dimensional quadrupole field. Wolfgang Paul, a 1989 Nobel laureate, invented it. It has a high resolving power for masses. It is used to confine ions in a small area, allowing for the simultaneous usage of MS and MS-MS analysis. Numerous ion trap devices have been created, including the three-dimensional ion trap, the linear ion trap (LIT), and the orbitrap. It employs oscillating electric fields to selectively capture ions. Figure 10 illustrates a typical ion trap mass analyzer. It comprises two hyperbolic ring electrodes and two electrically identical hyperbolic end cap electrodes. Utilizing a combination of RF and AC potentials applied to the ring and end cap electrodes, the following can be accomplished.

Trap all ions below a defined *m/z* range, all ions above a specified *m/z* value, all ions with a specified *m/z* value, and all ions with a specified *m/z* value. Eject ions with specified *m/z*

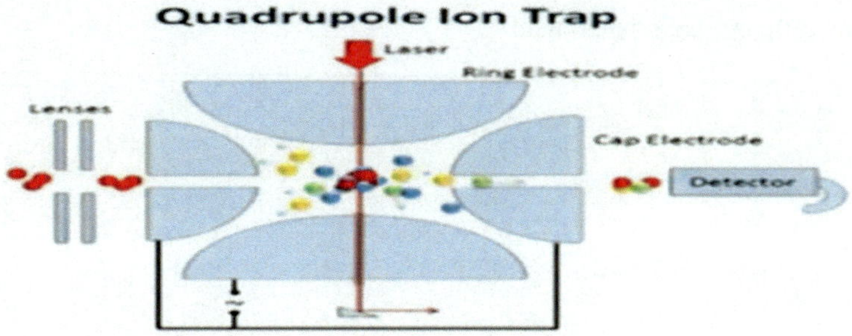

Fig. 10 Quadrupole ion trap mass analyzer [9]

ranges. These methods are advantageous for scientific purposes and can be applied qualitatively.

Advantages of using ion trap mass analyzer are that they are having high sensitivity, multiple product ion scan capability, high resolution. They are good for data dependent analysis (DDA) as it could predict which ions to be fragmented. But it also suffers from the drawbacks like production of unusual spectra if the ions are trapped for too long, it gets easily saturated, mass of less than 100 Da cannot be separated. It has a low dynamic range (except for the more recent devices) and therefore could have small quantitative applications.

4.1.5 Tandem MS

Different types of mass analyzer can be accommodated in one and called as tandem MS. Different mass analyzers are kept in tandem. It is done basically to fragment the precursor ions. Precursor ion means the ion which has been generated from a particular compound that has been ionized and subsequently these precursors ions are broken down in subsequent mass analyzers which gives daughter ions. They may be combined in different combinations like quadrupole-quadruple, magnet sector-quadrupole, quadruple-time of flight, time of flight-time of flight. In this, all ions are separated by fragmentation. When the molecule is entering into the mass spec it has to be broken down into smaller elements. Multiple method of fragmenting take place by various methods like collision induced dissociation (CID), electron capture dissociation (ECD), electron transfer dissociation (ETD), and chemically assisted fragmentation (CAF).

Table 1 presents the different types of experiments that can be performed in Tandem Quadrupole.

Q1 Scan—Q1 is in the scan mode, collision cell is inactive, and Q2 serves only as guide.

Q2 Scan—Q1 serves only as guide, collision cell is inactive, and Q2 is in the scan mode.

SIM (Single Ion Monitoring)—Q1 acts as mass filter. This system is operating as a single quadrupole MS system monitoring selected ions.

Table 1
Tandem Quadrupole experiments

Scan type	Q1	Collision cell	Q2	Data type
Q1 scan	Scan	Inactive	Rf only (acts as ion guide)	Qualitative
Q2 scan	Rf only (acts as ion guide)	Inactive	Scan	Qualitative
SIR	Mass filter	Inactive	Rf only (acts as ion guide)	Quantitative
Product ion (daughter) scan	Mass filter	Active	Scan	Qualitative
Precursor ion (parent) scan	Scan	Active	Mass filter	Usually qualitative
				Sometimes quantitative
Multiple reaction monitoring (MRM)	Mass filter	Active	Mass filter	Quantitative
Constant neutral loss	Scan	Active scan	Scan	Qualitative
Constant neutral gain	Scan	Active	Scan	Qualitative

Product Ion Scan—The system is set up to pass and fragment a specific ion and then monitor for its fragment ions (productions). It is primarily used for qualitative analysis.

Precursor Ion Scan—The system is set up to pass multiple ions (within the scan range) which are fragmented in collision cell and then a specific ion or set of ions are monitored in the third quadrupole. It is primarily used for qualitative analysis.

Constant Neutral Loss—This involves both Q1 and Q2 to be set for scanning across the whole *m/z* range, but both quadrupoles are off-set so that Q2 allows only those ions which differ by a certain number of mass units (equivalent to a neutral fragment) from the ions transmitted through Q1. This mode is used as a screening tool for classes of compounds like carboxylic acids tend to fragment by losing a (neutral) molecule of carbon dioxide, CO_2, which is equivalent to a loss of 44 Da or atomic mass units. All ions pass through Q1 and into the collision cell. The ions detected from the collision cell are those from which 44 Da have been lost.

Multiple Reaction Monitoring (MRM)—The system is set up for selectivity, allowing only a selected product ion to be fragmented and one fragment ion to be detected. Multiple MRMs can also be used, as well as several fragments from a specified product ion can be used for confirmation purposes.

5 Advantages of MS

Very sensitive in the analysis of most organic compounds.

Can differentiate between two or more compounds having same molecular weight.

Can identify the compounds even in an unresolved chromatographic peak by spectral data (molecular weight).

To summarize, the prepared samples were generated into small molecules which are then introduced into the source of the mass spectrometer. The source ionizes the molecules and subsequently this enters into the mass analyzer. In the mass analyzer itself, the masses of those spectra are determined and subsequently it is captured by the detector and the detector deconvolutes from the detector which is captured in the form of the mass spectra. There are inbuilt software in the equipment which deconvolute these mass spectra into different mass. All the mass spectra are then deconvoluted into mono isotropic spectrum and subsequently used for identification of the molecules. Data analysis in mass spectrometer generates data in the form of list of ions that are called as compounds which are detected in the order of various mass-to-charge ratio for proteomics purpose. These m by z needs to be transformed into all possible peptides. Data analysis involves the identification of all the compounds and their differential quantitation in two or more samples. So, when the spectra are generated, they show the fragmentation of the peptides which are used to determine the sequence of those peptides by a source algorithm and subsequently its structure is determined.

6 Acquisition Methods

There are some parameters in the ion source and mass analyzer which are to be optimized to control the behavior of ions along the ion path from the ion source to the mass analyzer to the detector and also to get the precursor and product ions for your target analyte. These parameters will be different for different analytes because for each analyte, there are specific product ions.

1. Ion spray voltage: Voltage is applied to the needle (electrode) that helps with nebulizer gas (N_2) and temperature to ionize the sample. It is applicable only for the electron spray ionization source and not be used in APCI. In case of APCI, we use nebulizer current to be applied to the corona discharge. This voltage depends on the flow rate and polarity of the compound. Higher is the flow rate of the solvent carrying the sample molecules coming from the HPLC, more ion spray voltage is to be applied. Hence the flow rate should be optimized for the

better ionization of the sample. Flow rate of the solvent also affects the spray stability and sensitivity.

2. Declustering potential (DP): Voltage is applied to the orifice (opening where ions enter the mass) for declustering of solvent cluster ions to minimize the number of solvent ions entering the mass and helps to transfer only analyte ions to the mass analyzer. It is an important parameter for developing any acquisition method. It should be optimized for each analyte as there will be a specific DP for any analyte. The higher the voltage the greater the amount of declustering but if DP is too high, unwanted fragmentation of the analyte may occur. Increasing the DP to more than optimum will also defragment the analyte ion. There will be specific DP for each analyte and ranged from (0-300 V) in the positive mode/(−300–0 V) in negative mode. So, every instrument has to be optimized for the DP.
3. Curtain gas: Counter flow of a high purity nitrogen gas that prevents neutral molecules, solvent droplets, and solid particles from entering the mass but only the analyte ions will be accelerated to the mass analyzer against gas by attractive potential gradient. Curtain ions will be applied in the interface between the plate and orifice to prevent the neutral molecules, solvent droplets, or particles from entering the mass analyzer.
4. CAD (collisionally activated dissociation) gas: Applied in collision cell to help for fragmentation of precursor ions to product ions which are different for each analyte.
5. Collision energy (CE): Amount of energy applied to the precursor ions for its conversion to the product ions.
6. Collision cell exit potential: To accelerate product ions out of the collision cell to be transferred to Q2 which is also the mass filter to transfer the product ions to the detector.
7. Dwell time (DT) is the time spent acquiring specific MRM transition during each cycle or the time to get the peak for one MRM. Longer dwell time will achieve better sensitivity and S/N ratio but we should consider the peak shape. Ideally, peak must be eluted 10–15 times to get an accurate measurement of its area.

$$\text{Dwell time}(s) = \frac{Pw(s)}{n \times \text{MRM}} - P_t$$

P_w (s) = peak width, n = number of points/peaks, MRM = Number of MRM transitions, P_t = pause time.

8. Cycle time: is the time spent on monitoring the target analytes including the pauses. We should also consider the pauses when we determine the cycle time.

References

1. Gandhi K, Kumar A, Sarkar P, Aghav A, Lal D (2018) MALDI-TOF MS: applications in dairy and related sectors. Res Rev 2(2):19–27
2. https://en.wikipedia.org/wiki/Electrospray_ionization
3. Dubey NK (2020) Metabolomics. In: Herbs and spices. IntechOpen
4. https://www.chem.pitt.edu/facilities/mass-spectrometry/mass-spectrometry-introduction
5. Honour J (2003) Benchtop mass spectrometry in clinical biochemistry. Ann Clin Biochem 40:628–638. https://doi.org/10.1258/000456303770367216
6. Cornish T, Bryden W (1999) Miniature time-of-flight mass spectrometer for a field-portable biodetection system. Johns Hopkins APL Tech Digest 20
7. Santoiemma G (2018) Recent methodologies for studying the soil organic matter. Appl Soil Ecol 123:546–550. https://doi.org/10.1016/j.apsoil.2017.09.011
8. https://www.wikiwand.com/en/Triple_quadrupole_mass_spectrometer
9. https://in.pinterest.com/pin/388224430356492821/
10. Gandhi K, Kumar A, Sarkar P, Aghav A, Lal D (2018) NALDI-TOF MS: applications in dairy and related sectors. Res Rev 2(2):28–36

Check for updates

Chapter 11

Atomic Absorption Spectroscopy and Flame Photometry

Abstract

Atomic absorption spectroscopy (AAS) is a technique which is employed for the determination of 70–80 elements both qualitatively and quantitatively. The methods for determination are categorized as absorption, emission, or fluorescence with the detection limit ranging to ppm levels. Atomic absorption spectroscopy (AAS) is a well-known and accepted technique for determining elements up to the trace (μg/ml) and ultra-trace (sub-μg/ml) levels with great accuracy and acceptable precision. In this chapter, basic principles of AAS and techniques for the processing of samples like dry ashing, wet digestion, and microwave digestion have been discussed. The design and working of various components of flame AAS have been covered. The performance of flame and graphite furnace AAS has been compared. Various practical considerations for analyzing a sample on AAS have been elaborated. Apart from this, various types of interferences like spectral and nonspectral interferences which occur in AAS have also been discussed. The principle of hydride generation AAS (HGAAS) and mercury cold vapor generation has also been explained. The basic principle, design, and working of inductively coupled plasma-mass spectrometry (ICP-MS) and inductively coupled plasma-atomic emission spectroscopy (ICP-AES) have also been covered. Flame photometry has been compared with AAS. Applications of AAS in medicine, biochemistry, toxicology, water and air analysis, and agricultural and food stuffs have also been covered. A standard protocol for the determination of iron content by atomic absorption spectrometric method in infant milk substitute and determination of sodium and potassium content in milk samples by flame photometry have also been provided.

Keywords Flame and graphite furnace AAS, Inductively coupled plasma-mass spectrometry (ICP-MS), Inductively coupled plasma-atomic emission spectroscopy (ICP-AES), Hydride generation AAS (HGAAS), Mercury cold vapor generation

1 Introduction

Atomic spectroscopy is a technique which is employed for the determination of near about 80 elements both qualitatively and quantitatively. The methods for determination are categorized as absorption, emission, or fluorescence with the detection limit ranging to ppm levels. Atomic absorption spectroscopy (AAS) is a well-known and accepted technique for determining elements up to the trace (μg/ml) and ultra-trace (sub-μg/ml) levels with great accuracy and acceptable precision. The technique was discovered by Walsh Alkemade and Melatz in the 1950s.

Kamal Gandhi et al., *Advanced Analytical Techniques in Dairy Chemistry*, Springer Protocols Handbooks,
https://doi.org/10.1007/978-1-0716-1940-7_11,

2 Basic Principles of AAS

By AAS we mean, the absorption of radiant energy by atoms. The absorption of the radiant energy is quantitatively related to the concentration of the metal ions present in a sample solution. In AAS, the concentration of the elements is determined by determining the intensity of the external radiation absorbed by the atoms of the sample at a wavelength specific for that element. The amount of the electromagnetic radiation absorbed by the atoms or ions in their gaseous state is measured in AAS. The principle involved is Beer–Lambert law which is as follows:

$$T = P/Po = e^{-\mathrm{kb}}$$

Absorbance, A, can also be expressed in terms of the transmittance as follows:

$$A = -\log\ T$$

$$A = abc \text{ or } A = E_o bc$$

$$a = \text{absorptivity in g/L-cm,}$$

$$E_o = \text{molar absorptivity in g/mol-cm}$$

$$b = \text{atom cell width in cm.}$$

Atomic absorption spectroscopy can be of flame absorption spectroscopy which utilizes the combination of different fuels and gasses like air/C_2H_2 (2300 °C) or N_2O/C_2H_2 (3000 °C). Flame absorption spectroscopy is less sensitive and requires high quantity of samples. Detection limit is in ppm or ppb levels. Graphite AAS is a type of AAS which utilizes graphite furnace as the accessory. The technique is highly sensitive (detection limit is in ppt levels), requires thermal pretreatment to the sample and lower quantity of the sample (even microlevels are sufficient). Other types of AAS are VGA/HG (vapor generation accessories/hydride generation accessories) which can be used for elements like As, Hg, Se, Bi, etc. which are hydride forming elements and can be analyzed at a very low level.

In atomic emission spectroscopy (AES), the radiation emitted by the atoms in their excited state by the virtue of heat or other means is measured. ICP-OES (inductively coupled plasma-optical emission spectroscopy) and MP-AES (microwave plasma-atomic emission spectroscopy) are other types of AES. In MP-AES, microwaves are used to generate the plasma and make use of the nitrogen gas. Nitrogen can be produced easily in the lab and is safe to use. Detection limits are at ppb levels. Temperature achieved is as high as 5500 °C. In ICP-MS (inductively coupled plasma-mass spectrometry), there is enough energy for the electron to leave the atom completely and leave behind a positively charged ion which is then

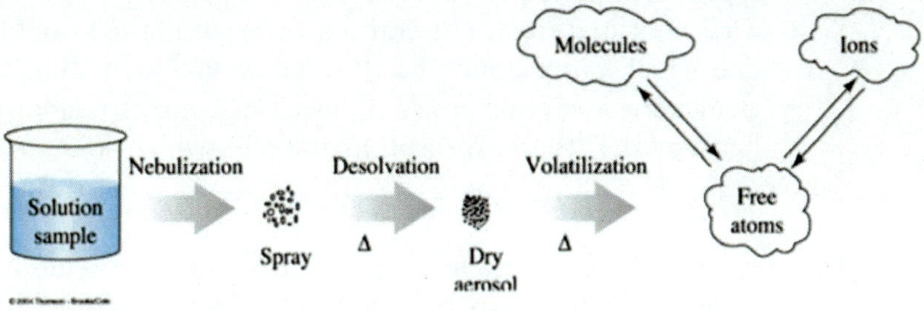

Fig. 1 Steps involved in the atomization of an element [2]

detected and quantified in mass detector. In ICP-OES and ICP-MS, temperature reaches to as high as 10,000 °C. ICP-OES is highly sensitive and fastest, while the ICP-MS is the most sensitive.

AAS mostly befits for analytical measurements as the atomic spectra generated contain discrete lines, with each spectra being unique for an element. The presence of other ions or atoms of different elements does not affect the identification and quantification of individual elements. The prerequisite for atomic spectroscopy is to convert the atoms or ions of the element of interest in its atomic state and this is achieved by atomization. Atomization facilitates the separation of the particles into the individual molecules through vaporization which further breaks the molecules into their atomic state (Fig. 1). The energy required for the conversion of an atom or ion into its atomic state is achieved by subjecting the element to heat in the form of flame or plasma.

2.1 Sample Atomization Steps [5]

(a) Nebulization: Process of converting the sample solution into fine droplets by passing it through a fine sized nozzle or through a vibrating crystal.

(b) Desolvation: The nebulized droplets of the sample are heated to evaporate the solvent leaving behind the analyte and its matrix compounds.

(c) Volatilization: Conversion of the solid analyte/mixture articles into gas phase.

(d) Dissociation: Dissociation of the molecules in gas phase to atoms.

(e) Ionization: Converts the atoms into charged moieties.

(f) Excitation: By light, heat, etc. for the measurement of the spectra.

The energy absorbed by the elements in their atomic or ionic state leads to the reduction in the signal from the source corresponding to the atomic absorption. The atoms or ions absorb energy in their ground state from a source of radiation and get excited to a higher state or energy level. These excited atomic species emit energy while returning to their original ground state or a lesser excited state. The emitted energy is in the form of a spectral line which corresponds to the atomic emission, thus producing characteristic atomic emission spectra. A direct relationship exists between the atomic absorption and AES which is given as follows:

$$M_{(\text{unexcited})} + h\nu \rightleftharpoons M*_{(\text{excited})}$$

Since the energy levels of the electrons revolving around the nucleus are fixed, so they absorb or emit radiation of discrete wavelength. This difference in the energy levels of both the state is directly related to the frequency of the absorbed radiation:

$$E_e - E_g = h\nu$$

where E_e = energy in excited state.

E_g = energy in ground state.

h = Planck's constant.

ν = frequency of the radiation.

$\nu = (E_e - E_g)/h$.

Or, since $c = \lambda\nu$

$$\lambda = hc/(E_e - E_g)$$

where c = speed of light.

λ = wavelength of the absorbed light.

The above mentioned equation shows that the ions or atoms in their atomic state absorb or emit light of a specific or fixed wavelength. Thus, this shows that every element has an exclusive set of allowed transitions which therefore lead to a distinct spectrum. Na (g) 3 s 3p and 3p 5 s as well as other transitions are possible at the correct photon energy of a transition.

3 Processing of Samples

1. *Dry ashing*: For dry or powdered food samples like milk powder, one gm sample is charred in the silica crucible and then transferred to the muffle furnace for 4 h maintained at 550 °C. The ash thus obtained is cooled and dissolved in dilute HCl (6 M), which is then used for mineral determination.

   (a) *Advantages*

   - Dry ashing is a convenient and versatile method for sample preparation.
   - This method allows the use of relatively large sample size.

- It minimizes the contamination due to reagents, since in this procedure only dilute acid is used to dissolve the ash.
- Minimum attention of the operator is required.

(b) *Disadvantages*

- The ashing is usually done between 400 and 600 °C; therefore, elements like Se, Pb, As, and Hg are lost through volatilization or by adsorption on the walls of the crucible. Other metals such as tin may form unsoluble refractory compounds during ashing.

2. *Wet digestion*:

Wet digestion procedure requires the use of strong oxidizing acid mixture of nitric, sulfuric, and perchloric acid. Generally for routine applications mixture of HNO_3 and $HClO_3$ is used. The use of mixtures containing H_2SO_4 is particularly useful when the sample containing fats are to be oxidized. The addition of H_2SO_4 increases speed of wet oxidation process as it raises the temperature of digestion. Even H_2SO_4 and HNO_3 mixture can be used, but complete digestion of fat may take hours. For this reason, fat is generally separated from the milk before digestion.

Triacid mixture (HNO_3: H_2SO_4: $HClO_4$–3:2:1). Extra care should be taken to avoid charring of samples when elements like As, Se, or Hg are to be determined.

(a) *Disadvantages*

- Chances of contamination are high as relatively large volumes of reagents are required to be added.
- It is suitable only for smaller size samples only.
- The strong acids which are used in the process are hazardous for the health of the operator, if the fumes are inhaled.
- Besides this, constant supervision of sample is required.

3. *Microwave digestion*:

Recently microwave heating methods are available which involve digestion of samples with oxidizing acids under pressure in closed vessels. Microwaves are electromagnetic energy that brings about molecular motion and thus heating by migration of ions and rotation of dipoles, without causing any change in molecular structures. In this, the sample is put in the closed container together with acid and decomposed by heating with microwaves.

(a) *Advantages*

- The digestion can be completed in shorter time when compared to heating on the conventional hot plates.

- Since samples are contained in closed vessels (low- or high-pressure Teflon tubes fitted with pressure relief values), losses due to volatilization are reduced considerably.
- Apart from this, lesser amount of acid is used for digestion, so reduced chances of contamination are there.

In atomic absorption spectroscopy, atomization is done by using the following two modes:

- Flame atomic absorption spectroscopy.
- Electrothermal (graphite furnace) atomic absorption spectroscopy.

4 Flame Atomic Absorption Spectroscopy

A hollow cathode lamp (HCL) serves as a source of radiation, the cathode of which is generally made up of the element of interest. The sample to be analyzed is converted into the fine droplets using a sprayer or nebulizer and is then subjected to the flame, which evaporates the solvent from the droplets leaving behind the solute. These particles vaporize and get decomposed into their atomic, ionic form. The radiation of light produced from the HCL is passed through the course of the flame containing the atomic vapor. The atomic or ionic form of the element or its electrons when absorb the radiation, results in the reduction of the intensity of the radiation coming from the source. A photomultiplier tube converts the signal of the radiant power received from the source into a related electrical current (Fig. 2). The absorption thus obtained from the element of interest in the sample is compared with that obtained from the standard solution of known concentration, thus the concentration of the element in the sample is determined.

The beam originated from the hollow cathode lamp after absorption by the gaseous atoms in the flame and that from the excited atoms in the flame travels to the monochromator. A chopper (disc with segments removed) is placed between the lamp and

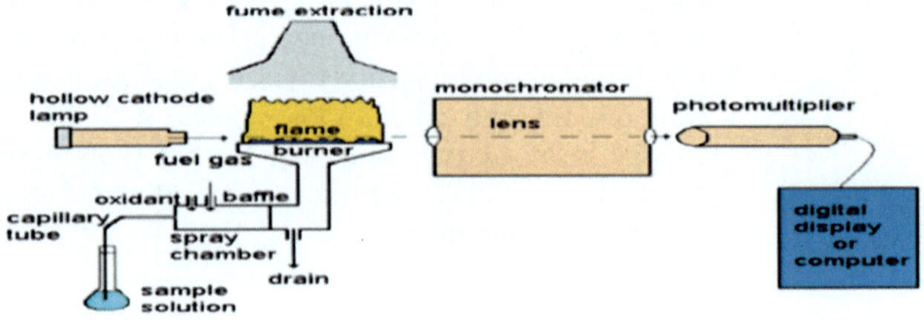

Fig. 2 Design of atomic absorption spectrophotometer [6]

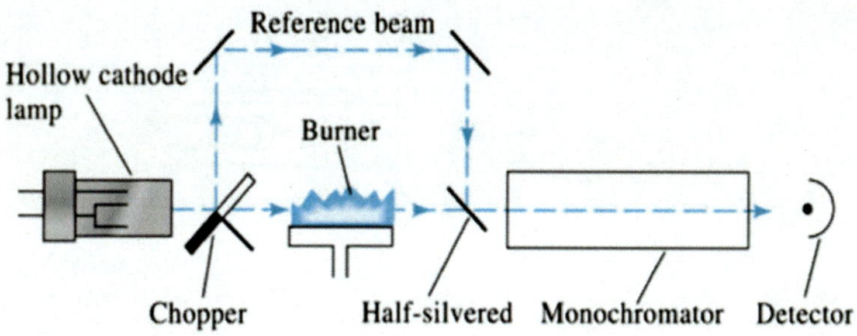

Fig. 3 Design of a double beam AAS

flame such that it is perpendicular to the light. The chopper rotates at a fixed speed in order to ensure that the beam of light reaches the flame alternatively at regular intervals (Fig. 3).

- AAS involves the striking of light of a particular wavelength onto previously generated ground-state atoms.
- The concentration of the atoms in the ground state is related to the intensity of the transition.

$$T = P/P_o$$

$T =$ transmittance.
$P =$ power of the light exiting from the sample.
$P_o =$ power of the light incident on to the sample.

$$T = P/P_o = e^{-kb}$$

In practice, a logarithmic relation is followed between the absorbance (A) and the transmittance (T), which is given as follows:

$$A = -\log T$$

The Beer–Lambert law relates to the concentration of the element in the atom cell, c as follows:

$$A = abc \text{ or } A = E_o bc$$

$$a = \text{absorptivity in g/L-cm,}$$

$$E_o = \text{molar absorptivity in g/mol-cm}$$

$$b = \text{atom cell width in cm.}$$

Nowadays with newer developments, the modern instruments can automatically convert the log value into A. Since the resonance transition (transition of highest intensity from the ground state or lower energy level to the first excited energy level) is most sensitive so it is widely used.

Flame AAS consists of five major components:

- Radiation (light) source,
- Atomizers,

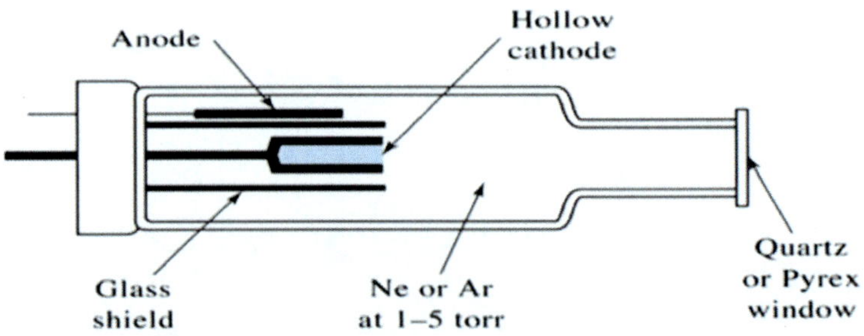

Fig. 4 Hollow cathode lamp [2]

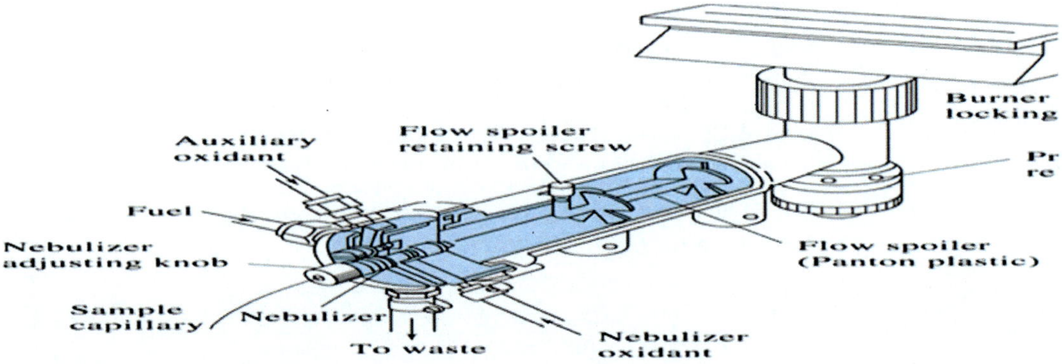

Fig. 5 Design of a flame atomizer [7]

- Monochromator,
- Detection system, and.
- Readout.

4.1 Radiation Source

HCL is the most commonly used light source. Single-element HCL contains a glass envelope filled with noble gas, like Ne, Ar, or He at a pressure of 1 to 5 torr. The electrodes used are generally made of tungsten (anode) while the element to be measured in its metallic form forms the cathode (Fig. 4).

A potential difference of 500 V is maintained between the two electrodes with the current being maintained between 2 and 30 mA. The filler gas gets ionized when the potential difference across the electrodes is applied as a result the ions of the gas move towards the respective electrode. The ions on striking the cathode lead to the removal of the metal ions from the cathode. On repeated collisions, the metal atoms get excited and produce spectra which are characteristic of the metal of interest on returning to their ground state.

4.2 Atomizers

The design of a flame atomizer is shown in Fig. 5. Its components include *nebulizer* and a *burner*. As discussed the sample is converted into fine droplets or aerosol on passing from the nebulizer. The most common type of nebulizers found are pneumatic nebulizers (PNs). PNs work by forcing a jet of compressed air, nebulization gas which aspirates and nebulizes the solution as it is sucked by a capillary tube in a chamber consisting of the oxidant and fuel. The baffles in the chamber remove the larger droplets of the sample, leaving a fine homogenously sized mist that mixes with the oxidant-fuel mixture and is carried to the flame.

4.3 Burner

The burner is a long, narrow slot capable of producing a 5–10 cm long flame. The type of the oxidant, fuel used can be manipulated along with the oxidant–fuel ratio to alter the flame characteristics as required. The combination of air–C_2H_2 as oxidant and fuel, respectively, is mostly used for flame generation in AAS. Other combination like N_2O–C_2H_2 as oxidant and fuel, respectively, is also used. The dissociation of the molecules into atoms is carried out by the flame produced.

The air–acetylene mixture which attains a temperature of 2400 °C converts around 40 to 50 elements into their ionic or atomic form rapidly and efficiently, while for some of the elements, a higher temperature is required which is achieved by using a mixture of nitrous oxide-acetylene flame (2800 °C) [2].

4.4 Monochromator

The primary objective of monochromator is to allow the wavelength (radiation) of interest arising from the radiation source and the element in the flame to reach the detector by removing the remaining wavelength(s). The monochromator lies in between the flame and detector. Glass or gelatine filters, prisms, gratings, etc. are used.

4.5 Detector

Photomultiplier tube (PMT), a photoelectric detector measures the intensity of the resonance line (with and without the sample in the flame) which is transmitted by the monochromator. PMT contains a photo emissive cathode and several dynodes in a vacuum. Each detector should have sensitivity, linearity, response time $<< 10^{-9}$ second, frequency dependence of response, and stability. PMT has the ability to convert the radiant energy into the corresponding electrical signal.

4.6 Readout Devices

The early AAS systems consisted of meters having calibrated scales as the readout devices. The newer models of the instrument contain digital displays or external computers.

4.7 Advantages of Flame AAS

- Technique is robust (foolproof and not affected by the presence of other elements present in solution).
- Easy to operate.

- Time-saving (few seconds/sample).
- Modest sample analysis cost/sample.

4.8 Disadvantages of Flame AAS

- The use of the toxic and flammable gases does not allow the unattended operation.
- Detection limits of the flame AAS fall in the range of 0.1–1.0 μg/ml range.
- A higher flame temperature required for refractory elements lowers the detection limits.

5 Graphite Furnace (Electrothermal) Atomic Absorption Spectroscopy (GFAAS)

GFAAS is similar to flame AAS except for the atomization process. GFAAS involves heating the sample to a temperature of around 2000–3000 °C which leads to the volatilization of the sample and causes its atomization. The signal generated by the furnace AAS reaches a peak in a few seconds. GFAAS is more sensitive than flame AAS, as it can detect minerals up to ppt levels. *Electrothermal atomizers* are graphite tubes cylindrical in shape to which an electrical power supply is connected. The sample is injected into the small hole in the tube with a microliter syringe. In order to avoid the tube from catching fire and for the removal of the air from the sample compartment, inert gas is flushed into the system. The injected sample is first evaporated by heating the tube electrically, the heated sample is then converted into its ash and the temperature is finally increased rapidly to 2000–3000 °C which vaporizes and atomizes the sample (Fig. 6).

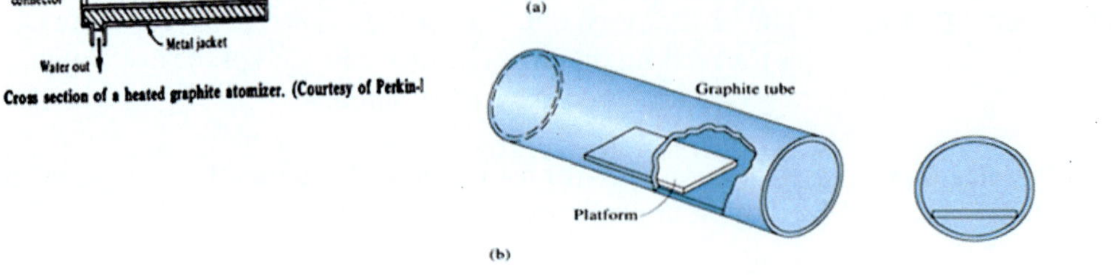

Fig. 6 Electrothermal atomizers

5.1 Advantages of Furnace AAS

- The amount or volume of the sample used is very less, that is, only microgram or micro liter.
- More sensitive than flame AAS (10 to 100 times).
- Toxic or flammable gases not used.
- Can operate unattended.

5.2 Disadvantages of Furnace AAS

- Analysis takes (typically) 2 min.
- Interferences arising due to chemical or physical factors require complex methods such as chemical modification, platform atomization, and background correction.
- Needs skilled and trained personnel to handle such a complex instrument.
- The added burden of lower sample throughput, lower precision, electrothermal furnace, and difficult operation increases the expenses.

5.3 General Practical Considerations of AAS

- *Reagents*:

  Food products generally contain majority of minerals at trace level or minute quantities. So, the chemicals, water used in the preparation of the sample, and standard solution should be highly pure.
- *Standards*:

  To quantify the level of the mineral in the sample, it is important to analyze a standard solution with a known concentration.

5.4 Sample Preparation

- Ashing of the sample is important to render it free from the organic matter followed by dissolution of the ash in a dilute acid (HCl) or water prior to analysis.
- In case of oils, they should be dissolved in an organic solvent like acetone or ethanol before subjecting the sample to the AAS flame.
- Milk samples should be freed from protein by precipitating it with TCA, thus facilitating the direct analysis of the supernatant obtained.

5.5 Labwares

- The glasswares used during the analysis should be free from any type of the element of interest, to prevent any cross-contamination.
- Use of plastic containers is recommended as glass has a higher affinity to absorb metal ions.

5.6 Calibration

- Calibration of the instrument is always necessary and should be done using appropriate standards.
- The absorbance obtained by analysis of a series of standard solutions can be plotted against their concentration.

5.7 Standard Addition Method

A solution of concentration x has an absorbance value D1 and its absorbance concentration curve comes to be a straight line. Similarly if a metal with a known concentration is added to it and its absorbance changes to D2, then x can be calculated from the relation:

$$x/x + a = D1/D2$$

The value of x can be calculated for this relation.

6 Interferences in Atomic Absorption Spectroscopy

The types of interferences encountered during the analysis of sample by AAS are discussed in the below sections:

6.1 Spectral Interference

- Absorption of source radiation.
- Nonspecific absorbance of the wavelength in the spectral band caused by the element present in the sample other than that of the interest, for example. The emission line for zinc, that is, 213.856, overlaps with that of iron, that is, 213.859.
- This interference can be resolved by reducing the slit width of the monochromator or by selecting an alternative emission line for the element of interest.

6.2 Absorption of Source Radiation by Background

The incomplete atomization of the sample may lead to the presence of particulate matter which may cause the scattering of the source radiation which leads to the attenuation of the radiation which reaches the detector. This type of interference can be resolved by ensuring the complete atomization of the sample by increasing the flame temperature.

6.3 Nonspectral Interference

6.3.1 Transport Interferences

- Occurs when something in the sample affects the rate of nebulization, transportation of the sample into the flame or aspiration.
- Other factors like surface tension, density of the sample solution, viscosity, vapor pressure also affect the rate of transport of sample into the flame.
- Such type of interference can be solved by matching the physical properties of the sample and standards as closely as possible.

6.3.2 Ionization Interference

Ionization of analyte atoms in the flame may cause a significant interference. Ionization leads to an equilibrium situation:

$$M \rightleftarrows M + +e^{-}$$

Elements like potassium which can be easily ionized can be used to suppress the ionization.

6.3.3 Solute Volatilization Interferences

This occurs when the element of interest reacts or combines with an interfering compound, resulting in the formation of a compound having a low volatility.

Presence of phosphate in the sample leads to the reduction in the absorbance caused by Ca.

This type of interference is overcome by

- adding another element, such as lanthanum (lanthanum diminishes interference due to phosphate and also acts as a releasing agent by facilitating the disassociation of calcium from proteins),
- using a higher temperature flame, and (describe what happen),
- adding a ligand such as EDTA, which can help in binding the analyte to form a complex, thus preventing its reaction with the interferant.

7 Hydride Generation AAS (HGAAS)

The hydride generation methods can be used to determine those elements which when made to react with sodium borohydride acid solution form volatile hydrides. Metalloid oxyanions on reacting with $NaBH_4$ and HCl produce volatile hydrides like H_3Sb, H_2Se, H_2Te, H_3As, etc [1]. The oxidation state of the metalloid plays an important role and should be taken care of before introducing the sample in the hydride generation system. Majority of the parts of the HGAAS system are identical with that of the AAS: HCL, air/C_2H_2 flame, and optical system which include an optical cell and the comparatively complex hydride generation system. Hydride generation is used for the elements that are difficult to analyze by flame or graphite furnace atomic absorption spectrometry. Elements with primary absorption lines below 250 nm suffer from interferences due to the absorption of the radiation by the flame and air. Hollow cathode lamps for these elements are usually low in intensity [1, 5].

Steps in hydride generation AAS (HGAAS)

1. Metal hydrides are generated by simple reaction chemistry.

   Acidified sample solution reacts with $NaBH_4$.

$$As^{3+} + 6\ BH_4^- + 3\ H^+ \rightarrow AsH_3(gas) + B_2H_6 + 3H_2$$

- Gaseous hydride is purged from solution by an inert gas.
- Carried to a silica gel (atomizer cell).
- Silica cell is heated by either flame or furnace.
- Hydride decomposes by either flame or furnace back to elemental form.
- Atomic absorption is measured.

Continuous sample and reagent mixing produce a steady state signal. Eliminates background interference and increases sensitivity.

8 Mercury Cold Vapor Generation

Mercury can be analyzed via cold vapor AAS. It has substantial vapor pressure at room temperature. There is not any requirement for heating. The acidified sample solution must be mixed with a reducing agent to generate elemental mercury. $NaBH_4$ or $SnCl_2$ can be used.

$$Hg^{2+} + Sn^{2+} \rightarrow Hg\ (vapor) + Sn^{4+}$$

Element mercury is carried to the spectrophotometer by carrier gas, where it is measured with a Hg hollow cathode lamp at 253.7 nm.

9 Correction of Drift

Drift is a change in the system over time that gives a positive or negative bias in the final result. It could come from any part of the system, including hollow cathode lamp radiation, alignment of optics, monochromator, or detector. Changes may be caused by the internal system or the surroundings. Drift can be corrected by using either double-beam optics or Stockdale optics [2].

9.1 Double-Beam Optics

Half of radiation from HCL passes through the atomizer, while rest of the radiation is used for the reference measurement. The ratio of sample beam/reference beam is then calculated. Corrects for drift in HCL radiation, optics, monochromator, and detector. It has a long-term stability. Sensitivity is reduced.

9.2 Stockdale Optics

All of the radiation from the HCL passes through the atomizer. Leaping mirrors direct radiation through the reference channel. The ratio sample beam/reference beam is then calculated. Corrects for drift in HCL radiation, optics, monochromator, and detector. It provides long-term stability, excellent detection limits, and excellent precision [2].

10 Correction of Background

Background are the nonspectral interferences that directly affect analyte sigma and are caused by the sample. It is not related to the drift which is caused by the instrument and the environment. Causes of background are:

- Scattering of radiation by particles in the atomizer.
- Absorption of hollow cathode radiation by molecules.
- Broad band generation.

Mostly in GFAAS but can affect flame.

The principle behind the background correction is to make two separate absorbance measurements [2].

First measurement with HCL gives total absorbance.

$$A_{\mathrm{HCL}} = A_{\mathrm{atomic}} + A_{\mathrm{background}}$$

Second measurement with a continuous source gives background absorbance only.

$$A_{\mathrm{broadband}} = A_{\mathrm{background}}$$

Subtracting second from first measurement gives corrected result.

$$A_{\mathrm{HCL}} - A_{\mathrm{broadband}} = A_{\mathrm{atomic}}$$

The sequence of measurements is repeated automatically. Background correction systems differ in the way the atomic signal is suppressed in the second measurement.

10.1 Advantages

- Well-proven.
- Highest sensitivity.
- Widest dynamic range.
- Cost-effective.
- Most widely used correction method.

10.2 Disadvantages

- Wavelengths restricted to UV (D_2 has low output in visible).
- Does not correct for spectral background.
- Does not correct for spectral overlap.
- Two sources are needed.

10.3 Structured Background

Continuum correction assumes heat background absorption as constant over full monochromator bandwidth. This is good assumption for most cases, for example, light scatter. However, some molecular absorbance bands show fine structure, for example, phosphate in 210–220 nm region Al salts in 190 nm region. We should measure background at exactly the same wavelength.

10.4 Zeeman Effect

Zeeman observed that gas atoms when exposed to a strong magnetic field $B \geq 1$ T experience splitting of electronic energy levels.

In the simplest splitting, one absorbance line generates two satellite sigma lines and one central π line. The π line is twice as intense as the sigma lines. The π lines absorb light polarized parallel to B. The sigma line absorbs light polarized perpendicular to B.

Principle of absorbance.

First measurement is taken when the polarizer is parallel to B.

$$A_{\text{parallel}} = A_{\pi} + A_{\text{background}}$$

Second measurement is taken when polarizer is perpendicular to B.

$$A_{\text{perpendicular}} = A_{\text{background}}$$

Molecular matrix absorbs regardless of polarization. The sigma lines absorb at slightly higher and lower wavelengths.

Subtracting second from first measurement gives corrected result.

$$A_{\text{parallel}} + A_{\text{perpendicular}} = A_{\text{atomic}}$$

10.5 Disadvantages of Zeeman Correction

For anomalous Zeeman elements, multiple central components introduce errors in background measurement

- Reduced corrected signal.
- Increased calibration curvature.
- Poorer S/N.
- Polarizer absorbs HCL radiation, so higher PMT gain must be used.
- Complexity and cost.
- Normally limited to atomization phase.
- Limited modulation frequency.

11 Inductively Coupled Plasma-Atomic Emission Spectroscopy (ICP-AES)

Plasma means a conducting gaseous mixture which contains a substantial concentration of ions and electrons. Argon plasma is commonly used in ICP where the principal conducting species are argon ion and electrons. Plasma torch contains three concentric quartz tubes through which the argon gas flows. The sample aerosol is fed via the central tube, the tangential argon plasma support flow helps cool the inner walls of the central tube. The top of the

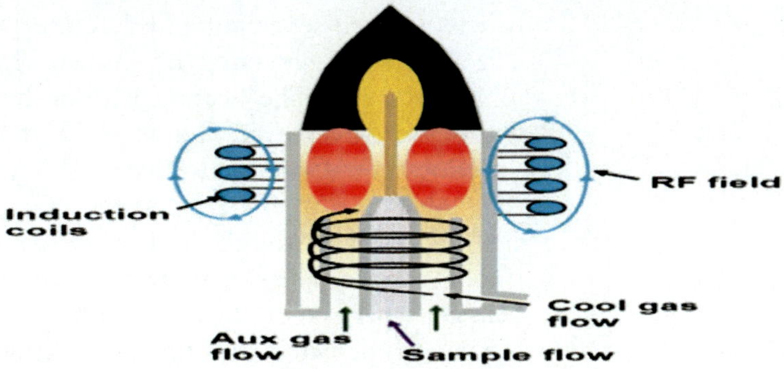

Fig. 7 Plasma torch From *Thermo Scientifica* [8]

outermost tube is enclosed by a water-cooled induction coil which runs by a radio frequency generator, which develops a variable magnetic field. A spark induced from the tesla coil initiates the ionization of the flowing argon gas. When the resulting ions and electrons flow through these regions, they interact with the fluctuating magnetic field. The interacting electrons and ions flow in annular path within the coil (Fig. 7). The resistance developed due to the flowing ions and electrons results in the ohmic heating of the plasma. Temperature of the plasma reaches as high as 6000–10,000 °C. The relatively uniform temperature in the plasma and relatively longer residence time give good linear responses over a several orders of magnitude range.

There are two types of ICP-AES instruments: sequential (monochromator) and simultaneous (polychromator) spectrophotometer.

Sequential ICP-AES instruments generally scan over a range of the wavelength such that the readings at several preselected wavelengths are made quickly but not concurrently. The scanning is accomplished by using a number of monochromators. This technique enables determination of a number of elements present within the sample during a single aspiration operation.

Simultaneous ICP-AES instruments monitor over several wavelengths concurrently. The polychromators present in the instrument are arranged such that they can isolate and focus on various spectral lines on a series of photomultiplier tubes which are arranged as a semicircle within the instrument. The advantage of these types of instruments is detection of various elements rapidly with outstanding precision but their cost limits their use and the wavelengths are preset to factory.

12 Inductively Coupled Plasma-Mass Spectrometry (ICP-MS)

ICP-MS detects low concentration of metals in the sample in which the MS is combined with a high-temperature ICP. The atoms of the metal or its salt in the sample are converted into the ions by the ICP which are then separated and detected by the mass spectrometer. The plasma is generated by inductively heating the gas which ionizes the gas. The heating is done by an electromagnetic coil which is capable of heating argon to more than 6000 K. Plasma contains neutral atoms, positive ions, free electrons, and molecules that form one of the four states of matter. The sample to be analyzed passes through the ICP. Due to the high temperature of the ICP, the sample gets ionized. The ions on entering the electric field get separated on the basis of their m/z ratio. Mass spectrum so obtained is depicted with the X-axis showing mass-to-charge ratio of the ions, while the Y-axis signifies the signal intensity. The signal intensities thus generated are directly related with the concentrations of the elements in the sample.

12.1 Introduction of Sample

With help of a peristaltic pump, sample is pulled in the machine through long, narrow pipe. Some amount of argon gas is mixed with it during the movement of the sample towards nebulizer. The sample moves into the nebulizer where they converted to fine spray called aerosol. In the nebulizer, argon molecule also gets converted into fine mist.

12.2 Working of RF Generator and ICP Torch

RF generator ionizes Ar molecule to positively charged ions. This argon molecule spray from a nebulizer in the fine mist form. This positive charged argon ions make plasma in ICP torch. Here the transfer of charge takes place between Ar ions and sample element.

12.3 Reaction

$$M + Ar^{+} \rightarrow M^{+} + Ar$$

This reaction takes place in ICP torch.

12.4 Monochromator

Now the charged sample ions travel towards the monochromator. This deviates all ions to their respective wavelength or in different colors of single wavelength. In some device, prism is also used.

The number of photons emitted from monochromator is proportional to the number of atoms that make it to the plasma.

12.5 Detector and Analysis Through Computer

Detector is used to detect all wavelength emitted by the monochromator. It converts wavelength into digital signal. So that a graph is plotted in computer. At last, qualitative and quantitative analyses take place. Table 1 presents the differences between ICP-AES and ICP-MS.

13 Flame Photometry

When a sample of an alkali or alkaline earth metal salt is shown in the Bunsen flame, a characteristic color in the flame is found

Table 1
ICP-AES vs ICP-MS

ICP-AES	ICP-MS
Separation of ions based on wavelength of ions	Separation of ions based on mass-to-charge ratio
Method development is simpler	Method development is quite difficult
Cost is low	Cost is higher
Detects 73 elements	Detects 82 elements
It detects 60 elements/min	Detects more than 60 elements/min
No highly specialist is required	Highly specialist chemist is required
It does not damage directly	Can be damaged by direct sunlight

Name of the element	Emitted wavelength range (nm)	Observed colour of the flame
Potassium (K)	766	Violet
Lithium (Li)	670	Red
Calcium (Ca)	622	Orange
Sodium (Na)	589	Yellow
Barium (Ba)	554	Lime green

Fig. 8 Colors produced by the metallic elements in the flame

(Fig. 8). This experiment is known as flame test and has been used for many qualitative analyses for metal ions and comes under atomic emission spectroscopy called flame atomic emission spectroscopy.

Most of the metals can be determined in a single solution without chemical separation. Flame analysis takes much less time than the corresponding chemical analysis; hence, it is suitable for routine analysis. Flame analysis is simple compared to chemical analysis. Amount of sample required is less than for chemical analysis. Accuracy is generally higher some time in fact better than chemical analysis. Preparation of sample is usually easier. This technique is used for the detection of alkali and alkaline earth metals like Na, K, Li, Ca, and Ba in environmental, food, clinical samples, etc. The sample solution containing the metal is aspirated into the

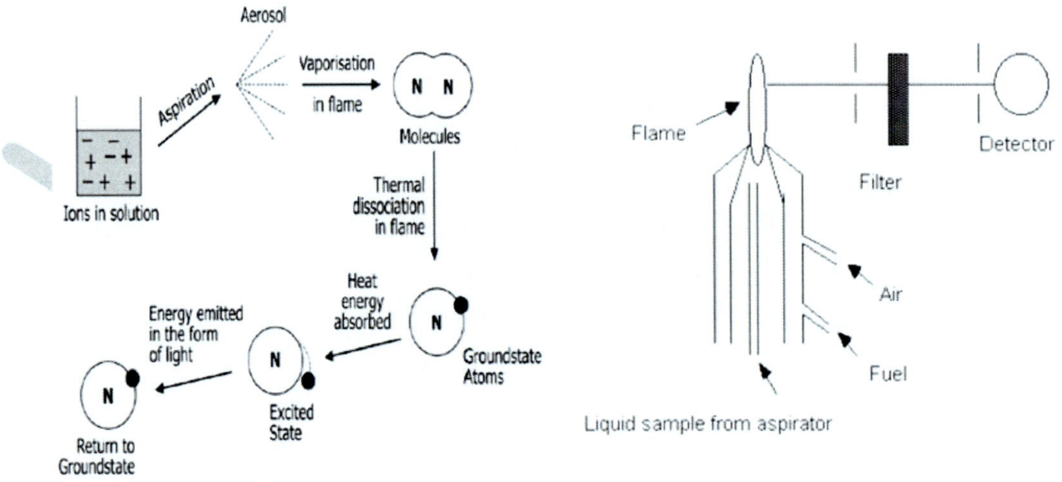

Fig. 9 Design of a flame photometer

flame. The flame helps in evaporation of the solvent, the metal atom gets atomized and under the influence of the external energy (flame) the valance electron of the metal gets excited to a higher energy level or upper state. On returning to its lower energy level or ground state the electron emits a characteristic wavelength (Fig. 9). Optical filters select the emission wavelength monitored for the analyte species. For quantitative analysis, the emission intensity of the sample solution is compared with that of a standard solution containing the known concentration of the metal or by using an internal standard. This method is suitable for the estimation of metals which can be easily ionized as the temperature of the gas and air flame used is lower than the other excitation modes like sparks, rare gas plasmas, and arcs. The temperature of the flame used is not high. This technique is only applicable for detection of alkali and alkaline earth metals and not for the transition metals (as the ionization energy required is very high). The common disadvantage related to this technique is that it is susceptible to interference or stability of both the aspiration and flame conditions. Factors like purity and fuel, oxidant flow rates, rate of aspiration, viscosity of the solution, presence of contaminants in the sample, etc. affect the flame and aspiration condition. Thus, the conditions maintained during the analysis of both the sample solution and standard solution should be as identical as possible.

Flame Photometer: The parts of the flame photometer include:

1. Burner.
2. Atomizer: Atomizes the solution into fine droplets or mist into the flame.

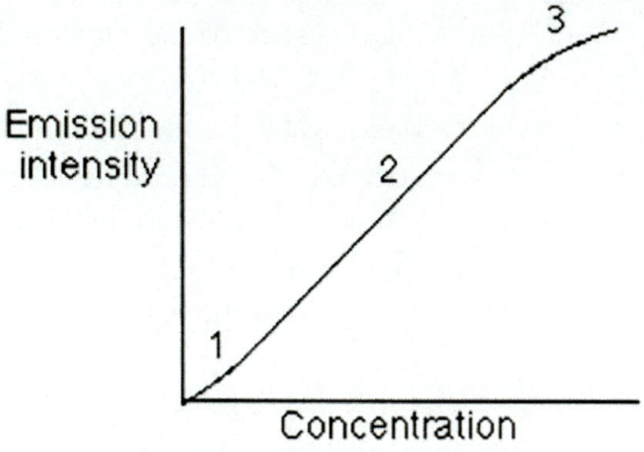

Fig. 10 Calibration curve for the estimation of metals

3. Filter system or diffraction grating: Isolates or filters the emitted light so that a specific wavelength characteristic to the element of interest reaches the detector.
4. Photocell: A detector cum converter. Converts the light energy into instrument friendly electrical signal (current).
5. Amplifier: For the amplification of the current.
6. Galvanometer: Measures the electric signal.

13.1 Estimation

Quantitative determination of an element present is usually done through the following methods:

In case of organic materials, namely food, milk, blood, etc. elements, namely sodium, potassium, etc. are determined through ash solution using flame photometer. Firstly the air pressure of the nonluminous flame is adjusted to 0.5 kg cm^2. Unknown ash solutions/diluted milk samples are then atomized in this range (on appropriate dilution of the concentrated ash solutions) and their concentrations are calculated from the standard curve (Fig. 10).

13.2 Application

Flame photometry represents an important standard method of modern analytical chemistry. Method is generally utilized for alkali and alkaline earth metals present in several substances. Biological fluids including blood, soil, plant materials, foods (milk and milk products), glasses, water are few substances wherefrom these elements could be easily determined both qualitatively and quantitatively using flame photometry.

13.3 Disadvantages

- Substance to be analyzed must be in the form of solution.
- Sometimes solution needs to be diluted to a considerable extent.
- Regular calibration by means of standard sample is necessary.

13.4 Preparation of Sample

- The sample should be free from large solid particles to prevent the blockage of the capillary because the sample is drawn by a very fine capillary.
- Biological fluids like milk or blood should be diluted before use.
- Samples of solid products like khoa, cheese, paneer, channa, etc. should be ashed and dissolve the ash in dilute acid before use.
- The sample should be diluted such that the concentration of the metal should be between 5 and 8 ppm.

13.5 Preparation of Standard Solution

The standard solution of the metal or its salt is the solution whose concentration is known. The concentration of such solution should lie between 5 and 10 ppm. If the concentration of the metal in the sample is very less, the organic solvent used for preparation of the standard and the sample should be the sample or the mixture of the organic solvent and water.

13.6 Procedure for Preparation of Stock

Standard solutions

1. Sodium.
   (a) 1000 ppm (as Na):
   Dissolve 2.539 sodium chloride in 1000 ml of glass distilled water.
2. Potassium.
   (b) 1000 ppm (as K):
   Dissolve 1.911 g KCL in 1000 ml of glass distilled water.

Standard solution is then diluted to give working solution of sodium as 10, 9, 8 ppm and potassium as 5, 4, 3 ppm, respectively.
1000 ppm solution for Na and K

$$10 \text{ ml } (1000 \text{ ppm}) \underset{(1:10)}{\rightarrow} 100 \text{ ml } (100 \text{ ppm})$$

$$10 \text{ ml Na and } 5 \text{ ml K} \rightarrow 100 \text{ ml } (10 \text{ ppm Na and } 5 \text{ ppm K})$$

$$9 \text{ ml Na and } 4.5 \text{ ml K} \rightarrow (10 \text{ ppm Na and } 4.5 \text{ ppm K})$$

Milk Sample

$$5 \text{ ml milk} \underset{(1:10)}{\rightarrow} 50 \text{ ml}$$

$$5 \text{ ml of this} \underset{(1:10)}{\rightarrow} 50 \text{ ml}$$

13.7 Determination of Metals in the Unknown

Firstly the instrument must be calibrated and standardized with a standard solution whose concentration is known. As the instrument gets adjusted, then the sample is placed in the flame and analyzed. The mineral content is recorded in ppm.

13.8 Precautions during the Analysis

- No solid particle above the size of the diameter of the capillary must be present in the sample. Pressure of burning gas and air must be optimum and there should not be any change in the pressure while feeding the standard and the unknown sample.
- Flame must be completely colorless when distilled water is fed to the instrument.
- Distilled water being used for the preparation of standard solution or sample must be free from the concerned metal being determined.

Sodium and potassium salts, for example, NaOH, KOH, Na_2CO_3, $NaHCO_3$ are used as neutralizers. Their presence in milk and milk products can be detected by determining the sodium and potassium content in them. The concentration of sodium in buffalo and cow milk ranges between 45 and 55 mg/100 ml and 50 and 60 mg/100 ml, respectively, while that of potassium is 100 and 120 mg/100 ml and 140 and 150 mg/100 ml in buffalo and cow milk, respectively. Concentration of these minerals above the normal range signifies the presence of added neutralizer in milk.

14 Flame Photometry vs AAS

Table 2 presents the differences between flame photometer and AAS.

Table 2
Comparison between flame photometer and AAS

Flame photometry	Atomic absorption spectroscopy
It is based on emission	It is based on absorption
The radiation emitted from the flame is measured	Decreased radiation intensity from the hollow cathode lamp due to absorption by the atoms in the flame is determined
Only alkali and alkali earth metals can be estimated	About 80 elements can be estimated
Filters isolate the emission line	Monochromators isolate the emission lines
HCL and chopper are not needed since the excited atoms or ions in the sample serve as a source of radiation	HCL and chopper are required for the source of radiation
Beer's law is not obeyed	Beer's law is obeyed
Any variation in the temperature affects the absorption intensity and signal response	Independent of temperature

15 Applications of These Techniques

Medicine, Biochemistry, and Toxicology

- Estimation of Ca and Mg in various body fluids and analysis of Na, K, Fe, Zn, and Cu in serum.
- AAS also serves ideally for the determination of the iron binding capacity and iron in hemoglobin using the graphite tube furnace.

Water and Air analysis

- The metal salts which are dissolved in water can be detected directly.
- AAS permitted the direct analysis of air for Pb and Cd vapors.

Agricultural and food stuffs

- Analysis of soil, soil extracts, fertilizers, and plants for Na, Ca, K, Mg, and trace elements like Cu, Mn, Fe, Mo, Zn, and B.
- Fruits, vegetables, fish, and meat products require either wet or dry ashing.

Dairy Industry

AAS is used to determine various minerals in milk and milk products. Singh (2019) [3, 4] assessed the contamination of the milk and milk products with heavy metals. In another study they also evaluated the profiling and distribution of minerals content in cow, buffalo, and goat milk.

15.1 Determination of Iron Content in Infant Milk Substitute: Atomic Absorption Spectrophotometric Method

Test type: Quantitative.

Reference: AOAC 999.11: 2016

Principle

The sample is firstly dried and then ashed in a muffle furnace at 450 °C ± 25 °C. After ashing, dilute (6 M) HCl is added and evaporated to dryness. The leftover residue is then dissolved in 0.1 M HNO_3 and used for analysis.

Apparatus

General: All the glasswares and plastic wares should be cleaned thoroughly with 10% HNO_3 and kept in it for atleast 6 h. Rinse them three times with double-distilled water before use to make them free from acid and allow to dry before use.

The cleaned glass and plastic ware should be stored in a dust free environment to prevent from any contamination. Laboratory apparatus, plasticwares and glasswares which are required for the analysis are as follows:

(a) Analytical balance.

(b) Volumetric flasks.

(c) Pipettes.

(d) Micropipette.

(e) Measuring cylinder.

(f) High-density polyethylene (HDPE) bottles, meant for storing the standard and sample solutions.

(g) Silica crucibles, of capacity between 30 ml or 50 ml.

(h) Programmable furnace oven, capable of reaching to a minimum temperature of 450 °C ± 25 °C.

(i) Flame atomic absorption spectrometer with an air–acetylene burner, appropriate for calculating at the wavelength of 248.3 nm for iron (Fe) ion content-determination procedures fitted with single element or hybrid hollow cathode lamps.

(j) Boiling water bath or hot plate.

Reagents

Analytical grade reagents should be used unless not specified.

(a) Deionized or redistilled water.

(b) Hydrochloric acid—6 M.

(c) Nitric acid (HNO_3), 65% (w/w).

(d) Nitric acid (HNO_3) solution, 0.1 M.

(e) Iron standard solution—1 mg/ml. Dissolve 1 g Fe in 14 ml water +7 ml nitric acid (c), in 1 L volumetric flask. Dilute to volume with water.

(f) Working standard solution—Dilute standard (e) with 0.1 M HNO_3 to a range of standard that covers the concentration of the element to be determined.

Preparation of test sample

There should not be any contamination of the test sample. Mix gently by inversion only.

Procedure

(a) *Test portion for dry ashing*

Weigh, to the nearest 1 mg, 10 g of prepared test sample in the previously dried silica crucible.

NOTE: If it is necessary to verify that the repeatability criterion is fulfilled, two different determinations under repeatability conditions should be carried out.

Dry ashing

The silica crucible should be placed in the furnace oven set at room temperature. The oven should be started as described: for drying and pre-ashing steps of the sample, increase the temperature by 50 °C/h up to 550 °C and hold for 6–8 h. When the obtained ash is of gray color, add 1 ml of HNO_3 to dissolve the ash.

Determination

(a) *Preparation of the test solution: Dry ashing: Dissolve the obtained ash in 1 ml of nitric acid solution. Transfer quantitatively the crucible content into a 250 ml one-mark volumetric flask by rinsing with water. Dilute to the 250 ml mark with water. Mix thoroughly and dilute the sample up to the required dilution.*

(b) *Dilution: According to the type of test sample and the ion measured, dilute (dilution factor, f1) the test solution by using the micropipette in the required one-mark volumetric flasks. Add a volume fraction of 10% (one tenth of the measuring flask volume) of the lanthanum trichloride solution by using a graduated measuring cylinder. Dilute to the mark of the volumetric flask used with water.*

(c) *Blank test: In parallel with the procedure of the test portion, carry out a blank test using the same procedure and the same amount of each reagent being added in the decomposition and the determination steps of the test portion.*

(d) *Flame atomic absorption spectrometric measurement: Adjust the flame spectrometer and its flame conditions according to the manufacturer's recommendations in order to yield optimum precision and sensitivity. Set the spectrometer at the required wavelength depending on the ion (analyte) to be determined.*

Calibration graphs

All the calibrating solutions and the zero solution should be analyzed at least thrice for calibrating the instrument. Calculate the average of the absorbance values. The difference between the absorbance value of the zero solution from the means of absorbance values will then be calculated. The obtained net absorbance should be plotted against the corresponding ion concentration.

Flask number	Iron ion solution (mg/l)
1	0
2	1.0
3	2.0

(continued)

Flask number	Iron ion solution (mg/l)
4	3.0
5	4.0
6	5.0

Measurement of test solution

The test solution and the blank should be immediately analyzed for their absorbance values once the instrument has been calibrated without altering the conditions of the instrument. The test solution should be diluted (dilution factor f2) using the zero solution if the obtained absorbance is higher than that of the highest standard. Add that much quantity of lanthanum trichloride solution to each dilution in order to achieve a final concentration of 10% solution. At least one of the calibrating solutions should be analyzed at the end of the measurement to check for any of the drift that occurred during measurement. The sample should be analyzed to get a minimum of three readings. Calculate the mean of the absorbance values. Calculate the difference in the mean value from that of the blank.

Use average corrected value to read the corresponding content from the calibration graph.

Calculation

Calculate the ion content, *w*, by using the following equation:

$$w = \left(\frac{(a - b) \times V}{m \times 1000}\right) \times f1 \times f2$$

where

w = is the ion mass fraction, expressed in milligram per gram, of the test sample.

a is concentration in test solution (mg/l).

b is mean concentration in the blank solutions (mg/l).

V = is the volume, in milliliters, of the flask in which the dry ashes (*V*=250 ml) or digested solutions (*V*=50 ml) are transferred quantitatively.

m = is the mass of test sample used in the procedure.

f1 = is the dilution factor of the test solution carried out during the preparation step.

f2 = is the dilution factor of the test solution carried out during the measurement step.

Principle *The estimation of sodium and potassium in various samples is generally done by flame photometry. Introduction of the sample into the nonluminous flame in the form of a mist causes the ionization*

15.2 Determination of Sodium and Potassium Content in Milk Samples by Flame Photometry

of the valance electrons of these metals as a result they absorb energy and get transferred to a higher energy level. On returning to their ground state or a lower energy level the electrons emit a characteristic wavelength. The intensity of the emitted light is directly proportional to the concentration of metal ions over a particular range of the wavelength. Instrument should be calibrated with a series of the standards before the start. The standard solutions should cover all the range of concentrations which is expected to be present in the sample.

Materials and reagents

1. Flame photometer.
2. Sodium stock solution (1000 ppm sodium): 0.254 g NaCl dissolved in 100 ml deionized distilled water.
3. Potassium stock solution (1000 ppm potassium): 0.191 g KCl dissolved in 100 ml of deionized distilled water.
4. Combined working standard solution of sodium and potassium ions. The concentration of working solution should be between 5 and 10 ppm.

Preparation of sample

The milk/skim milk sample to be analyzed should be diluted with deionized/distilled water to get the metal concentration in the range of 5–8 ppm.

Procedure

1. Follow the operator's manual instruction for running the flame photometer.
2. Firstly, the instrument is to be connected to an air compressor and gas cylinder.
3. Air compressor is then switched on and air pressure is adjusted through the instrument.
4. Outlet of the gas cylinder is then opened and the burner is ignited.
5. Standardize the instrument with standard solutions after setting the readout to nil using deionized/distilled water. Test for precision and reproducibility by several measurements of the standard. Make sure to aspirate deionized/distilled water between the measurements.
6. Diluted milk sample is then fed into the instrument and reading is reported in ppm.

Calculation

Sodium and potassium contents are then determined in milk by multiplying their reading (ppm) with the dilution factor.

Precautions

1. Deionized/double distilled water from quartz assembly should always be used.
2. Solutions are then mixed using glass spatulas and ensure that these are not coming in contact with the skin as they are released from the skin.
3. There should not be any solid particles in the solutions.

At the time of feeding the standard and the sample, the pressure of burning gas and air should stay steady.

References

1. https://www.chromacademy.com/
2. Nielsen SS (2017) Introduction to food analysis. In: Food analysis. Springer, Boston, MA, pp 3–16
3. Singh M, Sharma R, Ranvir S, Gandhi K, Mann B (2019) Assessment of contamination of milk and milk products with heavy metals. Indian J Dairy Sci 72(6):608–615
4. Singh M, Sharma R, Ranvir S, Gandhi K, Mann B (2019) Profiling and distribution of minerals content in cow, buffalo and goat milk. Indian J Dairy Sci 72(5):480–488
5. Wilson K, Walker J (2010) Principles and techniques of biochemistry and molecular biology. Cambridge University Press, Cambridge
6. http://delloyd.50megs.com/moreinfo/AA.html
7. http://faculty.sdmiramar.edu/fgarces/labmatters/instruments/aa/AAS_Theory/AASTheory.htm
8. https://www.thermofisher.com/in/en/home/industrial/spectroscopy-elemental-isotope-analysis/spectroscopy-elemental-isotope-analysis-learning-center/trace-elemental-analysis-tea-information/icp-oes-information/icp-oes-system-technologies.html